PLANT NUTRITION
An Introduction to Current Concepts

A. D. M. GLASS

The University of British Columbia

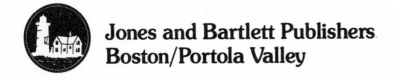

Jones and Bartlett Publishers
Boston/Portola Valley

Sales and customer service offices: 20 Park Plaza, Boston, MA 02116

Library of Congress Cataloging-in-Publication Data

Glass, A. D. M. (Anthony D. M.)
  Plant nutrition.

  1. Plants—Nutrition.   2. Plants, Effect of minerals on.   I. Title.
QK867.G575   1988   582.1'3   87-29863
ISBN 0-86720-080-4

Printed in the United States of America

Printing number (last digit) 10   9   8   7   6   5   4   3   2   1

Preface

Plant mineral nutrition is a subject of enormous scope. At the level of applied plant biology it is of critical importance in agriculture and forestry. The successful cultivation of many of our crops now depends upon increasingly sophisticated technologies and an intimate knowledge of plant physiology, particularly plant mineral nutrition.

At the same time plant mineral nutrition has a long history as a fundamental academic component of plant physiology and soil science programs; it is also essential for a proper understanding of plant ecology. Plant mineral nutrition currently draws from and interacts with developments in membrane biochemistry, biophysics and with cell physiology. It therefore offers immense challenge and opportunity.

This book is designed as an introduction to plant mineral nutrition, suitable for undergraduate students in botany, biology, soil science or agriculture as well as for graduate students and researchers. In discussing many contemporary topics, I have tried not to shy away from controversies. There are those who would argue that there is no place for elements of doubt in an introductory text. I disagree wholeheartedly; flourishing disciplines always generate contentious topics, lively debate and vigorous disagreements. It is important that students appreciate the excitement of such controversies. So I have included my opinions in some areas. These will no doubt be redundant within a short time, but that's how it is in science.

A.D.M. Glass
The University of British Columbia
Vancouver, B.C. Canada

Contents

*"Plant physiologists have two responsibilities
to the public whose money supports them.
One is to make profound discoveries.
The other is to make useful ones."*
J. B. PASSIOURA

1

Introduction

1.1 The Need for Useful Discoveries

Mankind presently numbers approximately 5.0×10^9 people, all of whom depend upon plant productivity for their sustenance. The task of providing adequate nutrition for such enormous numbers of people is one of staggering proportions, and, in the main, success has been achieved only in the developed nations. On a global basis the failure to feed mankind has as much to do with politics and food distribution as it has to do with plant nutrition. Regardless of these issues, the problem of feeding people in Third World countries is exacerbated by the massive rate of population growth. How did we arrive at the present situation?

Over the past million years human populations have grown in three massive waves (Fig. 1.1), each of which followed upon a critical technological advance. The first such advance was the development of tools and weapons about a million years ago. This enabled Paleolithic man to become a skilled and efficient hunter, but one that was, nevertheless, entirely dependent upon the provisions of a natural ecosystem. Figure 1.1 shows that this population plateaued at about 2 million people. The next great surge of population growth followed the emergence of the first primitive agricultural systems, which probably developed in the Middle East about 11,000 years ago. These

1

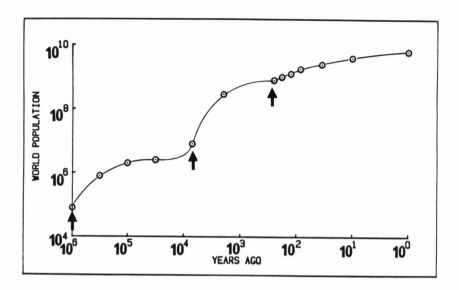

Figure 1.1. Growth of human population over the last million years. Note that time is plotted on a log scale. Times of major increases (shown by vertical arrows) correspond to the development of tools, agriculture and the industrial revolution, respectively.

allowed for much more efficient use of the available land. As techniques of cultivation advanced so did population numbers. Sophisticated civilizations became possible wherein the arts and the sciences could flourish. Finally, in the late eighteenth century the industrial revolution paved the way for the most recent rise of population. Since that time mankind's numbers have increased unabated, at such a pace that the world's present population is now predicted to double in a mere thirty-nine years!

In most of the developed nations population growth is no longer a major problem. Highly productive agricultures produce more than enough food for home consumption so that the excess can be sold abroad, earning valuable export dollars. Highly industrialized nations which need to import food can do so in exchange for their manufactured goods. Nevertheless, the agricultural industry is not without problems. For example, the current heavy dependence on fossil fuels makes the operation of Western styled agricultures an extremely costly enterprise. It is estimated that by the time we add together all the energy inputs for farm machinery, fertilizers and irrigation, as well as those required for harvesting, processing and packaging the crop, a total input equivalent to about six times the caloric value of the crop may have been expended.

During the 1970's the quadrupling of oil prices caused a massive ripple effect throughout the world economy, including the food industry. In the United States, retail food prices rose significantly faster than non-food prices during this period (8% per annum compared to 6.8%). By contrast, during the preceding decades (1950–1970) the situation was quite the reverse. Despite the preoccupation with energy in the 1970's, it is instructive to learn that in Canada, during this period, total farm expenditures on petroleum and other fuels increased only three fold, compared to a sixfold increase in expenditures on fertilizers. During the same period expenditures on petroleum, fuel and oil by U.S. farmers increased five fold (1.7 to 8.9 billion dollars) and dollars spent on fertilizers increased four fold (2.6 to 10.1 billions). As well as these absolute increases, the proportion of total farm expenditures accounted for by these two components increased from 9.2 percent in 1971 to 13.8 percent in 1981 (see Fig. 1.2). Current research reveals that any increases in farm operating costs pass to the

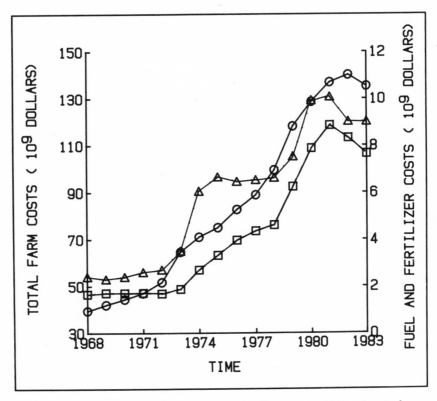

Figure 1.2. Total operating costs (O), fertilizer costs (△) and petroleum, fuel and oil costs (☐) for all U.S. farms for the period 1968-1983 in billions of dollars. (*Source*: United States Department of Agriculture Reports)

consumer, in the form of increased food prices, within six months.

Although crop yields have increased in recent years, the increases have been attained at the expense of ever more intensive (and hence more costly) farm management. This intensification has involved increased use of fertilizers, herbicides, pesticides and extensive irrigation. In turn these practices have led to agricultural losses through erosion and salinization of precious arable land. Such losses can lead to a need for even greater intensification of cultivation practices.

According to some assessments, the limits for further expansions of crop productivity by the above strategies may already have been reached. If we are to maintain present levels of production, let alone keep pace with the world's growing demand for food, there must be a massive increase of funding for research in all aspects of crop physiology. The recognition of the finite nature of petroleum supplies in the 1970's led to enormous increases of capital investment for the development of new and alternative sources of energy. It is forecast that in the 1990's and beyond, this situation will be repeated for the food industry. Factors limiting plant growth and crop yield will have to be identified and methods devised to circumvent these limitations. There will be exciting opportunities for plant nutritionists and for plant breeders, as well as for scientists who are skilled in the newly emerging areas of genetic engineering. Clearly, the successful resolution of these challenges will require the coordinated activities of research groups which are represented in all of the above areas. It is in the context of such a prospect that it behooves all plant physiologists to heed the advice of J.B. Passioura, to produce not only profound discoveries, but also useful ones.

Despite the gains in agricultural productivity in the developed countries the situation is still quite bleak for the underdeveloped nations. Vast numbers of people are malnourished and infant mortality remains high. As suggested earlier, the eradication of these problems will not be achieved through increases in plant productivity alone, for there are immense difficulties of a political, economic and cultural nature to be overcome. However, at least in the area of plant productivity there are opportunities for improvement, and indeed positive developments have occurred. Through the activities of individual research scientists and a worldwide network of international agricultural research centres, such as the International Rice Research Institute (IRRI) in the Philippines and the International Maize and Wheat Improvement Centre (CIMMYT) in Mexico, considerable progress is being made in the development of improved varieties of crop plants. The introduction of new high yielding crop varieties has been called the Green Revolution. The success of such programs depends on the integration of knowledge of all aspects of plant growth and nutrition, as well as agricultural technology. Of paramount importance among these scientific disciplines is the study of the absorption and utilization of inorganic nutrients, that is, plant nutrition.

1.2 The Need for Profound Discoveries

If the study of plant nutrition required any further justification we should consider the importance of nourishing the mind and the spirit of mankind. The means whereby plants take up and utilize the sixteen or so chemical elements which they require represent fascinating and challenging problems. These problems have occupied scientists for over three centuries.

The elements required by plants must commonly be procured from soils in which their concentrations are extremely low. Under such conditions the competition for the scarce inorganic resources may be intense. Nevertheless, the required elements are obtained and frequently concentrated within plant tissues to levels which are thousands of times greater than their environmental distributions. Two examples may serve to illustrate the point.

Using six day-old plants, grown in hydroponic containers, we measured the absorption of K^+ from 70 liters of a 5 μmolar KCl solution, radioactively labeled with ^{42}K. Despite the extremely low concentration of K^+ (5 μmol 1^{-1} is equivalent to 2×10^{-7} g of K^+ cm^{-3} or 1 K^+ per 10^7 molecules of water) the roots of these young plants (whose internal K concentration was 20 mM or 1 K^+ per 4×10^4 water molecules) reduced ambient K to undetectable concentrations within a few hours. Each gram of roots had effectively processed 3000 cm^3 of solution, removing essentially all available K^+ from solution against a 4000 fold concentration gradient. By six weeks each barley plant had achieved a dry matter content of 5 g, composed principally of carbohydrates. In order to obtain sufficient CO_2 to generate this dry matter, from an atmosphere containing 0.034% CO_2 (by weight), the leaves of each plant must have processed the equivalent of 20,000 liters of air!

If the scarcity of a particular element poses a challenge for survival, the opposite situation, namely excess availability, may be even worse. Fortunately the latter situation is not as common as the former, but where such situations occur they provide valuable insights into physiology, ecology and evolution. For example, in serpentine soils, so called because the parent rock (serpentine) is often mottled like a snake's skin, Mg is present in such excess and Ca is so scarce that most plants are unable to survive. As a result, plant diversity is quite low. However, this provides an additional bonus to those plants which have evolved mechanisms to deal with the soil imbalance. Not only are new territories made available to the successful plants but, in addition, the harmful effects of soil chemistry serve to screen out potential competitors. *Helianthus bolanderi* is a species that is endemic to serpentine soils in California. It grows successfully at Ca levels which are so low that its relative, *H. annuus* (the sunflower), simply expires. Krukeberg, at the University of California, showed that there were distinct races (or ecotypes) of *Gilia capitata* growing on serpentine soils which could also grow on non-serpentine soil when experimentally transplanted. However, the ecotypes which normally grow on non-serpentine soils grew very poorly when transplanted into the serpentine soils.

As well as toxicity problems arising from excesses of the required elements there are many similar examples for the non-essential elements. Under arid conditions some alkaline soils contain toxic quantities of selenium (Se) in the form of selenate (SeO_4^{2-}). So characteristic are the species which have colonized such seleniferous soils that they are referred to as indicator species. Some mining companies refer to the technique of locating mineral deposits by means of indicator plants as biogeochemical prospecting. *Astragalus* species (the milk vetches) are known to accumulate thousands of parts per million (ppm) Se without detrimental effects. Animals which feed upon plants growing in such areas may suffer selenium toxicity, the infamous 'alkali disease'. More details of this subject matter are discussed in Chapter 9 in the section on acidic and saline soils.

In the coming chapters we will explore many current aspects of plant nutrition. Although physiologists ultimately aspire to describe whole plant function, it is often necessary to discuss the component parts before attempting to put the whole plant in perspective. Chapter 2 examines the development of current concepts in plant nutrition while Chapter 3 describes the chemical and physical distributions of the nutrient elements in soils, the principal media for supplying inorganic nourishment to plants. In Chapter 4 we explore the way that root morphology and anatomy have evolved to meet the challenges posed by this distribution.

Chapters 5 and 6 discuss current perceptions of the mechanisms of ion absorption by roots and the transport of the absorbed nutrients through the plant. In Chapter 7 the effects of environmental variables on ion absorption are explored, as well as the physiological adaptations which enable plants to accommodate to these changes, while Chapter 8 deals with the biological functions of the elements in plants.

Chapter 9 concludes our introduction to plant nutrition by considering genotypic variations in nutrient acquisition and utilization, and in tolerance to toxic soil situations.

Summary

We are rapidly approaching the time when further increases in food production may no longer be obtained by the empirical methods of the past. In this coming era there will be exciting challenges, of both an intellectual and humanitarian nature, awaiting the plant nutritionist, who will probably form part of a research group providing essential information for the food industry. Already, many of the large chemical companies are investing considerable sums of money in areas such as plant breeding, tissue culture, and the genetic engineering of crop plants. A vital component of the advance from empirical to conceptual methods in the food industry will be a clear understanding of the physiology and biochemistry of plant nutrition.

Further Reading

Brown, L. 1976. Man, food and environmental interrelationships. In *Nutrition and Agricultural Development*, eds. N.S. Scrimshaw, M. Behar, pp. 3-12. New York: Plenum.

Deevey, E.S. 1960. The human population. *Scientific American* 203: 194-204.

Van Overbeck, J. 1976. Plant physiology and the human ecosystem. *Annual Reviews of Plant Physiol.* 27:1-17.

"Those who cannot remember the past are condemned to repeat it"
GEORGE SANTAYANA

2

The Development of Current Concepts of Plant Nutrition

2.1 Introduction

George Santayana's warning is as appropriate for an aspiring scientist as it is for the politician or economist. The scientist who focusses exclusively upon contemporary research, while neglecting the historical roots of his subject, does so at the risk of rediscovering the wheel. More importantly, in the present context there is no better way for the student to develop powers of criticism, judgement and discernment, while simultaneously learning the subject matter, than by studying the experiments and interpretations of our scientific predecessors. In this chapter we consider the major questions which appear to have motivated researchers of plant nutrition in recent history. We trace the gradual emergence of answers to these questions and examine how they led to contemporary concepts of plant nutrition. Subsequent chapters elaborate upon these fundamental concepts.

2.2a THE MAJOR QUESTIONS WHICH HAVE MOTIVATED PLANT NUTRITIONISTS

1. What are the sources of plant nourishment; soil, water, air? (Chapters 2,3)

9

2. Which of the chemical elements found within plant tissues are essential for growth and development? (Clearly such a question could not even be formulated prior to the great advancements in chemistry which occurred in the late 18th century). (Chapters 2 and 8)

3. What are the functions of the essential elements and what are the consequences of deficiencies of these elements? (Chapter 8)

4. How do the elemental compositions of plants compare with those of the habitats (terrestrial or aquatic) in which they normally grow? (Chapters 3, 5 and 7)

5. What special adaptations enable plants to procure the essential elements? In land plants, whose roots may be considerable distances from the photosynthetic organs (leaves), how are the elements supplied to aerial organs? (Chapters 4, 5 and 6)

6. How are rates of absorption and internal concentrations of the essential elements influenced by external (environmental) variables such as temperature, light and nutrient concentrations? (Chapter 7)

7. To what extents are plant growth and crop yield dependent upon the rates of provision of the major (fertilizer) elements? How can responses to these inputs be optimized? (Chapter 9)

2.2b   THE MAJOR CONCEPTS OF CONTEMPORARY PLANT NUTRITION
(and answers to the questions asked in 2.2a)

1. Terrestrial plants obtain their nourishment from simple inorganic resources; C, H and O from the air and from water, and the remaining elements as inorganic ions e.g. Ca^{2+} or oxides e.g. NO_3^- from soil solution. Plants which grow in lakes and oceans receive the essential elements from the aqueous habitats in which they are found. The free energy required to elaborate these inputs into complex biochemicals is provided by photosynthesis.

2. All plants have an absolute requirement for at least sixteen chemical elements.

3. Each of these essential chemical elements performs a specific biochemical or biophysical function within plant cells. Hence, deficiency of even one of these elements can impair metabolism and interrupt normal development.

4. The chemical composition of plant cells is typically quite different, both qualitatively and quantitatively, from the soils and aquatic systems in which they grow. Consequently there is a need

for highly selective accumulating and excluding (active transport) mechanisms.

5. Plants have evolved morphological, physiological and ecological mechanisms to ensure that they obtain sufficient quantities of the essential elements. Nutrients absorbed by the roots are delivered to the xylem and transported to the aerial parts in the ascending sap.

6. Many external variables can influence rates of ion absorption and accumulation. Nevertheless, the concentrations of the essential elements are maintained within rather narrow limits in plant cells. Such constancy is thought to arise from the operation of delicate feedback systems, which enable plants to respond in a homeostatic fashion to environmental fluctuations.

7. Despite the operation of the above homeostats, when external conditions severely limit the acquisition of the essential elements growth rates and tissue concentrations decline and deficiency symptoms may develop. As the availability of the essential elements is increased tissue concentrations and growth rates increase at first and then tend to saturate. Further increments of growth demand extremely high nutrient inputs.

2.3   The Identification of the Essential Elements

The list of sixteen chemical elements required by plants to sustain healthy growth and development includes C, H, O, N, S, P, K, Ca, Mg, Fe, B, Mn, Cu, Zn, Mo and Cl. This list was not completed until 1954 when Broyer and his colleagues at the University of California added Cl to the existing group of fifteen. Even now we should be cautious in presuming that the list is complete. Additional requirements for specific plant groups are still being discovered. For example, in the 1960's, Brownell, at the University of Queensland (Australia) established that Na was essential for certain species of *Astragalus*. As recently as November of 1983 it was reported, in the journal *Science*, that Ni is essential for legumes and possibly for all higher plants.

In order to achieve a definitive demonstration that a particular element is essential for plant growth it is necessary to be able to grow the experimental plants in rigorously defined media, using chemicals of the highest purity. In the 1860's, using the method of hydroponic culture, von Sach and Knop established that ten chemical elements were absolutely essential for plant life. These elements, referred to as the macronutrients because they are required in relatively large quantities, included C, H, O, N, S, P, K, Ca, Mg and Fe.

The period in which the essential role of these elements was discovered was one of immense change. The impact of science was being felt throughout the whole of society, notably in the agricultural sector. The population increases which arose from the Industrial Revolution and the exodus of manpower from the country to the growing industrial centres, demanded greater and greater efficiency of the agricultural community. The realization of the importance of the chemical elements for plant nutrition was soon followed by the development of a fertilizer industry which played a major role in enabling farming sectors to meet these challenges. Some indication of the developments which occurred in agriculture in this period can be obtained by examining wheat yields in Western Europe. From Roman times to about 1800, yields had changed but little, remaining at about 6-10 bushels per acre. By 1900 typical yields were three times these values. Figure 2.1 demonstrates that historical trends in the yields of

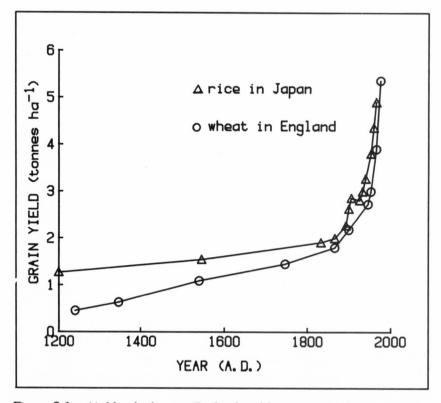

Figure 2.1.   Yields of wheat in England and brown rice in Japan over the last several centuries. (From Evans, *American Scientist* 60:388-397, 1980)

brown rice, in Japan, during this period were quite similar.

Prior to the changes which were initiated in the 1800's, agricultural methods had remained essentially unchanged for many centuries, based upon an ancient oral and traditional knowledge. Such practices as crop rotation, fallowing, liming, the use of green and animal manures, as well as the planting of leguminous crops to improve soil fertility, were already well established in Roman times. It is hard to believe that such techniques might have arisen without careful observations, some trial and error and even some systematic study. However, the origins of these techniques are lost in the antiquity of agriculture and, despite their widespread practice in Medieval Europe, there was, generally, little understanding of the underlying scientific principles.

Throughout the middle ages the powerful influence of Aristotle's doctrines of plant nutrition prevailed without challenge. Aristotle (384-322 B.C.) perpetuated the ancient belief that all matter was composed of the four basic elements; earth, air, fire and water. In essence he believed that the soil represented the sole source of nourishment for plants, and that "the roots of plants are analogous to the mouths of animals, both serving for the absorption of food". He considered plant food, like that of animals, to be in an already elaborated form, "which is the reason that plants produce no excrement, the earth and its heat serving them in the stead of a stomach".

Van Helmont, a Belgian physician, undertook his now famous experiments (published in 1648) to test the Aristotelian dogma. He placed 200 pounds of dried soil in a clay pot. The soil was moistened with rain water and a willow tree was planted in the soil. After 5 years, during which the soil was regularly watered with rain water, the willow grew from 5 to 169 pounds in weight. On drying and reweighing the soil, a mere two ounces appeared to have been lost. Van Helmont concluded that the plant substance had come entirely from water, and that the loss of weight of the soil was attributable to weighing error. Although this conclusion was different in detail from Aristotle's belief, it was not at variance with the fundamental concept of the composition of matter in terms of the basic elements earth, air, fire and water. Van Helmont believed that many diverse forms of matter, "Grisle, Wood, Barke and Leaves" are destroyed by burning and "do straightway of their own accord, give their part to water". Was it unreasonable, then, for him to accept that water might generate plant matter by processes which were the reverse of those he observed during burning?

The experiment conducted by van Helmont heralded the beginning of the experimental period of plant nutrition. In addition, it illustrates some important issues about science in general. We must be conscious that our interpretation of experimental results is powerfully influenced by prevailing concepts. We may chuckle condescendingly at van Helmont's naivete, but seen in the context of his time his conclusions were quite appropriate. I have little doubt that we are making equivalent errors today, channeled in our rationalizations by the dogma of our times. It is important, also, to appreciate that a hypothesis or mathematical model which adequately describes the phenomenon at hand may only be one of several satisfactory models. Compatibility between observation and hypothesis does not constitute proof of the hypothesis. As more information becomes available about the system under study, more complete, more refined hypotheses will inevitably emerge. The original hypothesis may have to be discarded altogether. Living systems are exceedingly complex and our attempts to probe their workings often lead us along false trails and blind alleys.

In 1656 Glauber (a German chemist) isolated saltpetre (KNO_3) from cattle manure. He correctly argued that since the cattle feed upon plants, the latter must represent the original source of this compound. Glauber demonstrated that saltpetre strongly stimulated plant growth but mistakenly presumed that it was the sole plant nutrient. In the same period, Palissy had recognized that dung returns to the soil what the plant removes during its growth. He stressed, moreover, that the salty ash which remains when a plant is burned, contains the salts which the plant removes from the soil. This interpretation was confirmed by Mayow's measurements of the nitrate concentration of the soil at different seasons of the year. He noted the decline of concentration as the growing season progressed and concluded that this was due to the nitrate being "sucked out (of the soil) by the plants". In 1699 Woodward published the results of his experiments with spearmint and other plants which he grew in rain water, river water, drain water from a Hyde Park conduit, and the same drain water shaken with garden soil. The results of one such experiment are shown in Table 2.1. Spearmint plants grown in a solution of

Table 2.1.   Woodward's Experiments with Spearmint

Solutions Provided	Increase of Plant Weight
(a) rain (= distilled) water	62%
(b) river water	93%
(c) drain water	126%
(d) drain water plus soil solution	309%

garden soil increased their weights far more than did the other plants. Since all plants had received adequate water, the differential growth in the different treatments must have been due to the absorption of "certain peculiar terrestrial matter".

Beginning in 1755, Home (an English chemist) made use of pot experiments to investigate the effects of various inorganic salts and some organic compounds (including olive oil!) on plant growth. Home noted the stimulatory effects of potassium nitrate, magnesium sulfate and potassium sulfate. Thus, it became evident, contrary to Glauber's belief, that at least four elements were required for plant nourishment.

In 1804 de Saussure published his "Recherches Chimique sur la Vegetation", and established a precedent for quantitative plant nutrition. He extended the approach of Woodward, growing plants in solutions containing single salts and even organic compounds. As a result of these investigations, de Saussure became convinced that the normal nutrition of plants was impossible without the absorption of nitrates and other minerals from the soil, even though these elements represented but a small proportion of the plant's food. In addition he observed that the absorption of the different substances required by plants was independent of their availability. From the analyses of plant ashes however, he was able to show that plant composition varied with the nature of the soil and the age of the plant. Further, he concluded that the phosphates and alkali contained in the ash must have been absorbed from the soil (it had been widely held that plants were capable of generating alkali, and that the acids contained in plant tissues were synthesized to neutralize these alkalis).

Despite this clear evidence of the importance of inorganic nutrients, von Thaer and Davy, among others, continued to perpetuate the notion that plants obtained their carbon from soil humus — the so-called "humus theory". Inorganic elements were considered to act merely as "stimulants". Davy even went so far as to advocate the use of oil as a fertilizer since it was rich in both carbon and hydrogen!

The humus theory was finally rejected after the publication in 1840 of 'Organic Chemistry in its Application to Agriculture and Physiology' by the German chemist Liebig. The author's considerable reputation, together with his fierce criticism of opponents, were important factors in the acceptance of his ideas. However Liebig mistakenly concluded that the source of plant nitrogen was atmospheric ammonia (NH_3).

The mid-nineteenth century saw further application of de Saussure's exemplary quantitative methods for the analysis of fertilizer effects in field trials, most notably by Boussingault at Alsace and by

Lawes and Gilbert in England. Liebig's notions regarding the source of nitrogen for plants were shown to be incorrect. For non-legumes, nitrates and ammonium salts, absorbed from the soil by the roots, satisfied nitrogen requirements. The experiments of these workers established without doubt that 'artificial' fertilizers could sustain soil fertility indefinitely.

Nevertheless, the numerous cases in which applied fertilizers failed to bring about increased plant growth were sufficient justification for more rigorous experiments. In the 1860's, von Sach and Knop reintroduced the methods of water culture, in which nutrient sources could be accurately defined. This led to the conclusive demonstration of an absolute requirement for the 10 macronutrients. Not until the early part of the twentieth century was it appreciated that the supposedly pure chemicals employed to prepare their culture solutions contained sufficient quantities of impurities to satisfy the needs for other (micronutrient) elements. Through the application of more exacting techniques to eliminate impurities in these solutions, the original list of ten essential elements has been extended to include B, Mn, Cu, Zn, Mo and Cl. Table 2.2 summarizes important historical information concerning the essential elements.

2.4   The Biological Roles of the Essential Elements

Although it was appreciated relatively early on that particular chemical elements were essential for plant growth, it has taken much longer to demonstrate the specific biological roles of each of these elements. Indeed, even today the functions of some of the required elements have not been clarified. In the cases of elements which form part of the chemical composition of plant tissues, such as C, H, and O in carbohydrates, the problem of ascribing function was relatively straightforward. As early as 1784, Lavoisier's appreciated that C and H were constituents of all organic matter. De Saussure's studies (1804) established that the elements of water were incorporated into organic matter with the "fixing" of carbon dioxide. Nitrogen had been established as a constituent of all plant proteins as early as 1785 by Berthellot.

However, for the remaining elements, many of which served more elusive physiological or biochemical roles, a clearer understanding of their function had to await the elucidation of the details of biochemical pathways and cellular physiology which occurred in the early part of the 20th Century. This is not to suggest that some of the earlier

Table 2.2.   A list of the essential elements, with the names of their discoverers and the dates of discoveries. Also shown are the names of the scientists responsible for proving definitively that these elements were essential for plants, with the dates of their critical publications.*

Element	Discoverer	Year	Discoverer of Essentiality	Year
C	**	**	De Saussure	1804
H	Cavendish	1766	De Saussure	1804
O	Priestley	1774	De Saussure	1804
N	Rutherford	1772	De Saussure	1804
P	Brand	1669	Ville	1860
S	**	**	von Sachs, Knop	1860's
K	Davy	1807	Lucanus	1865
Ca	Davy	1807	von Sachs, Knop	1860's
Mg	Davy	1808	von Sachs, Knop	1860's
Fe	**	**	von Sachs, Knop	1860's
Mn	Scheele	1774	McHargue	1922
Cu	**	**	Sommer	1931
			Lipman & MacKinnon	1931
Zn	**	**	Sommer & Lipman	1926
Mo	Hzelm	1782	Arnon & Stout	1939
B	Gay Lussac & Thenard	1808	Sommer & Lipman	1926
Cl	Scheele	1774	Broyer *et al.*	1954

* Some text books may give earlier authors but these may not have provided definitive proof of essentiality or may have demonstrated essentiality in but one species of plant.

** Element known since ancient times.

scientists were deterred from speculation. Text books and reviews of this subject written around the beginning of the present century make it evident that, with the exception of C,H,O, N and S, the roles of the remaining elements could be described in only the vaguest of terms. Calcium, for example, is deemed necessary for "the formation of the green parts of plants" and for "rendering harmless, toxic acids". Potassium salts are essential "for assimilation" (photosynthesis). Even as recently as 1966, a major review of the function of the elements written by Evans and Sorger (in the Annual Reviews of Plant Physiology) made it very apparent that there were still outstanding deficiencies in our knowledge of this important area.

The reasons for this situation are not hard to appreciate. Carefully controlled experiments can readily provide a positive or negative answer to the question of essentiality. The symptoms of deficiency, however, provide only the most obscure clues to function. For example, Nobbé's observations on the relationships between carbohydrate levels of tissues and K provision led him to conclude, in 1870, that K was essential for the biosynthesis of carbohydrates. However, it was not until 1968 that K^+ was shown to be essential for the activity of the enzyme starch synthetase. Ultimately, a definitive statement regarding function can rarely be obtained without details of the molecular basis of function. Chapter 8 is devoted to a detailed consideration of our contemporary understanding of the roles of the essential elements.

2.5   Selective Transport Systems

As a clearer picture emerged of the nutrient requirements for plant growth, it became possible to consider the mechanisms responsible for the absorption of these nutrients. Although Liebig had considered that plants absorbed substances in solution "exactly like a sponge imbibes a liquid, and all that it contains, *without selection*", the earlier studies of de Saussure had already given clear evidence of selective absorption by plant roots. Similarly the recognition by Nageli (1855) that plant protoplasts could behave like osmometers, swelling or shrinking according to the concentration of the external medium, provided strong evidence for the selective permeability of the cell to certain solutes. However, it was Pfeffer who placed membrane phenomena on a sure foundation by his use of artificial membranes of copper ferrocyanide to study osmosis and related phenomena. Pfeffer correctly concluded that the selective properties of the protoplast resided in the plasmamembrane and the tonoplast. In his text, "Osmotic Studies" (1871) he demonstrated rare insight, suggesting that the permeation of the plasmamembrane depended not only on the dimensions of the solute molecules and the membrane pores through which they were to pass, but also upon the capacity of the permeating molecules to interact (dissolve or react reversibly) with the membrane.

In 1886, Pfeffer used various basic dyes to explore the characteristics of membrane permeation, noting that low temperature or treatment with narcotics failed to impede the entry of these dyes. However, the topic of cell permeability was advanced in a systematic fashion, in the 1890's through the outstanding research of Overton who investigated the permeability of plant and animal cells to over 500 different substances. Overton was impressed by the universal similarity of permeability properties in tissues of widely different sources. He explored the relationships between chemical and physical characteristics of the permeating solutes and their capacity to penetrate the protoplast, establishing that the more lipophilic (fat soluble) the molecule, the more rapidly it penetrated the cell membrane (see Fig. 2.2). From this observation Overton argued that the cell membrane itself must be lipid in nature. Overton also stressed that although many solutes could enter cells passively, by virtue of their solubility in the cell membrane, many other substances were accumulated actively often from dilute external solutions. Unfortunately, the enthusiasm with which Overton's permeability studies were greeted somewhat obscured his equally important recognition of these active transport mechanisms and considerable misunderstanding and controversy

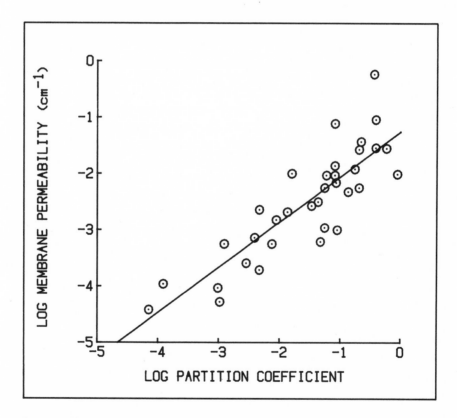

Figure 2.2   The relationships between lipid solubilities (represented by partition coefficients) of various compounds and their rates of entry into cells of the giant alga *Chara*. (From Collander and Barlund, *Acta Botanica Fennica*, 11:1-14, 1933)

regarding his conclusions ensued. As a consequence of this misinterpretation it was widely held that the entry of solutes into plant cells could be described entirely on the basis of diffusion forces and permeability properties. The misconception was not resolved until the 1920's by one of the most outstanding plant physiologists of the twentieth century. In 1914, Hoagland was appointed to the Berkeley campus of the University of California as an agricultural chemist. Since 1860 Germany had been the major supplier of the world's demand for K fertilizer. However, the disruption of trade which followed the commencement of World War I, interrupted America's supply of potash. The use of this element had climbed steadily, from 90,000 tons per annum in 1900 to 250,000 tons per annum by 1914. Clearly, there was an urgent need to locate alternative supplies of this element. Hoagland undertook a study of the giant kelps, which grow in the Californian waters, as a potential source of K. Impressed by the capacity of the kelps to concentrate ions such as K^+, Hoagland went on to investigate many aspects of ion transport. In studying Br^- accumulation by cells of the giant alga *Nitella*, Hoagland and his associates demonstrated that accumulation strongly depended upon the intensity and duration of the light source; in darkness there was little accumulation. These observations suggested that the absorption of Br^- depended upon a source of metabolic energy, namely photosynthesis. Soon after, also at Berkeley, Steward was to investigate K^+ and Br^- uptake by thin slices of potato tubers. As in the *Nitella* cells, significant rates of ion absorption were only obtained when conditions favourable to metabolism were applied; in this case, adequate aeration of the solutions in order to sustain respiration. These studies restored the emphasis to active processes as responsible for the absorption of nutrients. In the intervening years the major goal in studies of ion transport has been to identify the links between metabolism (respiration or photosynthesis) and ion absorption.

2.6   The Regulation of Ion Transport

Since Hoagland's clear demonstration of the linkage between metabolic energy and ion absorption, most of the research in ion transport has been directed toward clarifying the mechanisms (particularly the energy coupling) of these active transports. By contrast, only a relatively small group of researchers have been involved in studying the regulation of ion transport. Although a source of energy is essential to bring about active transport, considerations of energy alone would not allow us to anticipate the extremely delicate control of ion fluxes which enable plants to maintain internal concentrations of the elements at more or less constant values. Under natural conditions numerous environmental variables, notably temperature, light intensity, and ion availability would be expected to cause considerable variations in rates of ion absorption, tissue concentrations and growth rates. However, plants appear to respond homeostatically to these environmental disturbances so that internal constancy is maintained. This extremely important area has been under investigation since 1904 when it was first reported that plants starved of potassium developed an increased appetite for this element. Hoagland and Broyer (1936) enlarged upon these studies, but only recently has the area attracted renewed interest. The subject of responses to environmental factors is treated in detail in Chapter 7.

Summary

The recorded history of scientific research in plant nutrition spans little more than 300 years since the experiments of van Helmont, published in 1648. Nevertheless, there existed well organized and systematic agricultural practices for thousands of years before this time. It is inconceivable that such practices might have arisen without careful studies and observations of plant nutrition. Notwithstanding these, it was not until the 20th century that the list of 16 essential elements was completed and the fundamental concepts of plant nutrition were developed. The quest for an understanding of plant nutrition is not yet complete, however. The coming chapters will reveal that immense challenges still remain. Future developments in plant nutrition should prove to be as exciting as those of the past 300 years.

Further Reading

Bould, C.W. 1963. Mineral nutrition of plants in soils and in culture media. In *Plant Physiology: A Treatise. Vol III: Mineral Nutrition*, ed. F.C. Steward, pp. 16-96. New York: Academic Press.

Collander, R. 1959. Cell membranes: their resistance to penetration and their capacity for transport. In *Plant physiology: A Treatise. Vol II: Plants in relation to water and solutes*, ed. F.C. Steward, pp. 3-102. New York: Academic Press.

Krikorian, A.D. 1968. Water and solutes in plant nutrition. *BioScience* 18: 286-92.

Krikorian, A.D. 1975. Excerpts from the history of plant physiology and development. In *Historical and Current Aspects of Plant Physiology*, ed. P.J. Davies, pp. 9-97. New York State College of Agriculture and Life Sciences.

Reed, H.S. 1907. The value of certain nutritive elements to the plant cell. *Annals of Botany* 21: 501-543.

Russel, E.J., Revised by Russell, E.W. 1950. *Soil Conditions and Plant Growth*. 8th ed. London: Longmans Green.

3

Soils and Solutions: Media for Plant Growth

The plant kingdom comprises approximately 350,000 species, of which the majority are found on land. Moreover, although 75 percent of earth's surface is covered by water, terrestrial plants still represent mankind's principle sources of food as well as valuable forest resources. While it it true that, in recent years, commercial growers have made increasing use of hydroponic facilities to grow crops such as tomatoes, lettuce and cucumber (see Fig. 3.1), the vast majority of cultivated and wild plants obtain their inorganic requirements from soil. Therefore, the main emphasis in this chapter will be upon this medium, particularly its chemical and physical properties, in so far as these properties influence the provision of the inorganic elements to land plants.

3.1 Composition and Origin of Soils

Soils are exceedingly complex systems which are highly variable from region to region but universally consist of various proportions of three phases; solid, liquid and gas. In addition, soils generally contain

Figure 3.1.   A commercial lettuce crop grown in sand, irrigated with an inorganic nutrient solution. (Hidroponias Venezolanus, S.A., Caracas, in Resh, Howard M., *Hydroponic Food Production.* Santa Barbara, CA: Woodbridge Press, 1982)

a diverse community of interdependent plants, animals and micro-organisms. The solid phase contains the major inorganic reserves of the soil, while the aqueous phase (the soil solution) represents the immediate source of nutrients for absorption by plant roots. The gaseous phase, consisting of air-filled channels and pores through the soil, permits exchange of gases, particularly O_2, CO_2 and N_2, between the soil and the layer of air immediately above the soil. When these air spaces become water-logged for long periods, most plants are unable to survive.

3.1a   THE SOLID PHASE OF SOILS

The solid phase of most soils is made up of varying proportions of mineral matter, which is derived from underlying bedrock, and organic matter which results from the decomposition of plant material. On the basis of their mineral and organic matter content, soils may be classified as mineral or organic. Mineral soils generally contain but a small (5-10) percentage by volume of organic matter while the organic soils of swamps and bogs may contain as much as 80-95% of organic matter. The mineral soils are by far the more common. A good agricultural soil will generally contain 45% mineral matter, 5% organic matter and approximately equal proportions (25% each) of water and air by volume.

Mineral matter is derived from the parent bedrock by both mechanical and chemical weathering action (Fig. 3.2). Extremes of temperature cause cracking and breakup of the rock. Erosion by wind, water and ice continues this mechanical disintegration. Particles derived in this way without chemical change from the original magma are termed primary minerals. Quartz and silicate clays are examples of this group. Chemical decomposition may involve hydrolysis, hydration, oxidation and other forms of chemical change. Good examples of such changes include the hydrolysis of microcline, (a primary mineral consisting of $KAlSi_3O_8$):

$$KAlSi_3O_8 + HOH \rightarrow HAlSi_3O_8 + KOH$$
$$2HAlSi_3O_8 + 8HOH \rightarrow Al_2O_3 \cdot 3H_2O + 6H_2SiO_3$$

or the solubilization of calcium carbonate by conversion to the bicarbonate under the influence of rain water (which contains significant quantities of carbonic acid):

$$CaCO_3 + H_2CO_3 \rightarrow Ca(HCO_3)_2$$

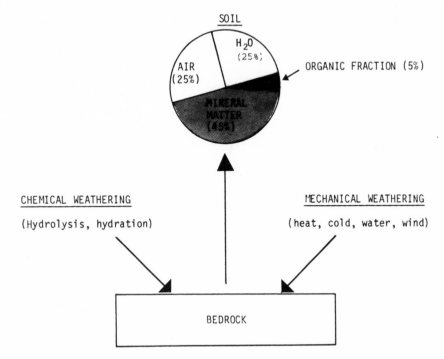

Figure 3.2.   Origins of the major components of a typical agricultural soil. Values shown represent percentages by volume.

Compounds arising from such chemical changes constitute secondary minerals.

Mineral particles of various sizes are generated as a result of the weathering process. The International Society of Soil Science defines clay particles as those below 0.002 mm in size. Silt particles (0.002 to 0.02 mm) sand (0.02 to 2.0 mm) and gravel (>2 mm) make up the remaining categories (Fig. 3.3). Each size group makes important contributions to soil character. The larger particles have little water retaining capacity but have the advantage of facilitating drainage and soil aeration. The smaller particles have physical properties dominated by their large surface area to volume ratio. Colloidal clay, for example, has about 10^4 times the surface area as the same weight of medium sized sand.

Chemically speaking the average composition of the soil of a particular area will be determined primarily by the chemical composition of the bedrock. Figure 3.4 shows the average elemental composition of the earth's crust (the lithosphere) to a depth of 16 km in terms of percentage by weight. Oxygen, Si and Al, in the form of various oxides (notably Si oxides) and aluminosilicates, clearly dominate (>80%) elemental composition. The nutrient cations Fe, Ca, K, Mg are relatively well represented, although (as in the case of Fe) the element may not necessarily be in a form which is available for plant absorption. By contrast, the inorganic anions, particularly those of N, S and P, and the micronutrient elements are relatively poorly represented.

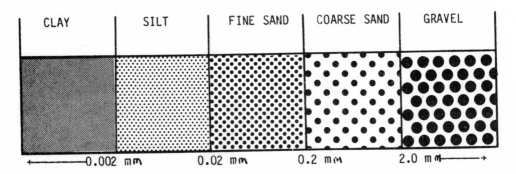

Figure 3.3. Soil mineral particles classified according to size. This classification is based upon the system proposed by the International Society of Soil Science.

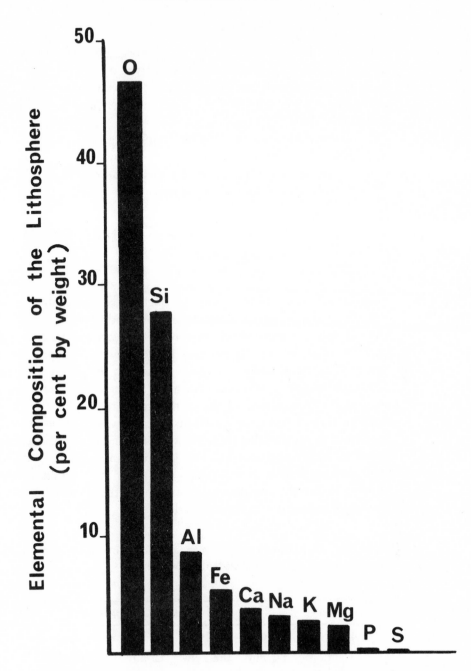

Figure 3.4. Elemental composition of the lithosphere expressed as percentage by weight. (From Mason, *Principles of Geochemistry*, John Wiley and Sons, 1958)

3.1b SOIL SOLUTION

A more meaningful description of the inorganic nutrients which are directly available for plant absorption is given by soil solution strength, even though this may vary considerably according to region, time of year, temperature, rainfall and numerous other factors. Despite these variations, Table 3.1 reveals a pattern which is fairly representative. Cations are well represented, often being present in millimolar concentration. However, ammonium is usually low due to its adsorption to soil particles and nitrification (the oxidation of NH_4^+ to NO_3^- by soil bacteria), which occurs in two steps:

(1)
$$2NH_4^+ + 3O_2 \xrightarrow{\textit{Nitrosomonas} \text{ bacteria}} 2HNO_2 + 2H_2O + 2H^+$$

(2)
$$2HNO_2 + O_2 \xrightarrow{\textit{Nitrobacter} \text{ bacteria}} 2NO_3^- + 2H^+$$

Nevertheless, since both groups of bacteria are obligately aerobic, NH_4^+ may accumulate in waterlogged soils and in organic soils where low pH and anaerobic conditions inhibit the activity of these bacteria. The other striking observation is the universally low concentration of phosphate, which is usually present in the micromolar range. Although the soil solution concentration is a better guide to availability than the composition of the earth's crust, the rate of turnover (or replacement) of soil solution from soil reserves may be even more important in determining total plant uptake and productivity. Experiments in which nutrients are provided continuously in hydroponic facilities have indicated that plants can attain high potential growth rates at extremely low concentrations of nutrients so long as these low levels are continuously replaced.

Table 3.1.   Typical concentration ranges for the major inorganic nutrients in soil solution and aquatic systems.

	Nutrient Concentration				
Environment	NO_3^-	Pi	K^+	Ca^+	Mg^+
1. soil solution	1-5 mM	0.2-2 μM	0.5-5 mM	1-10 mM	1-10 mM
2. marine waters	0-40 μM	0-2.5 μM	10 mM	10 mM	50 mM
3. fresh waters	0-40 μM	0-2 μM	50-150 μM	0.25-2 mM	1 mM

3.1c THE GASEOUS PHASE OF SOILS

Adequate gas exchange between soil and air is normally essential for plant growth. However we should not underestimate the enormous variation among plants and the existence of specific adaptations enabling certain plants to deal with the problems associated with poor aeration. Adequate aeration is important because of the high respiratory rates of plant roots. The active absorption of nutrients and the continued 'exploration' of soil demands a steady supply of energy from oxidative respiration. When oxygen levels decline, as in water-logged soils, glycolysis may represent the only source of ATP. Under these conditions pyruvate is converted to ethanol:

$$CH_3\text{-}\underset{\underset{O}{\|}}{C}\text{-}COOH \longrightarrow CH_3CHO + CO_2 \xrightarrow{\quad NADH+H^+ \qquad NAD^+ \quad} CH_3CH_2OH$$

(pyruvate) 　　　　　 (acetaldehyde) 　　　　　　　　　 (ethanol)

This compound is quite toxic; even 1% solutions can inhibit root growth. However, anaerobic conditions also lead to the formation of toxic end products, e.g acetic and butyric acids, by anaerobic micro-organisms. The reducing conditions which prevail in the absence of oxygen favor microbial reduction of NO_3^- to N_2 and N_2O. Eventually, more reduced elements such as Mn^{3+} and Fe^{3+} ions serve as oxidants and these are reduced to Mn^{2+} and Fe^{2+} forms. In some species e.g. *Mercurialis perennis* (Dog's mercury), the presence of excessive levels of Fe^{2+} in waterlogged soils appears to be responsible for root death, for when the species was grown in water logged sand, from which Fe^{2+} was excluded, the plants grew quite healthily. As progressively stronger reducing conditions prevail SO_4^{2-} is reduced to S^- and H_2S and eventually even H_2 and methane may be produced. Nevertheless, certain plants, e.g. rice and some aquatic angiosperms, possess morphological and physiological adaptations which enable atmospheric oxygen to be transported internally to submerged tissues and even released to the soil. Typically, such species possess an extensive system of air spaces between parenchyma cells. Such an arrangement, termed aerenchyma, permits the transfer of oxygen to submerged plant parts. Figure 3.5 shows such aerenchymatous tissues in the stem of *Nymphaea* (water lily) and in *Juncus* (rush).

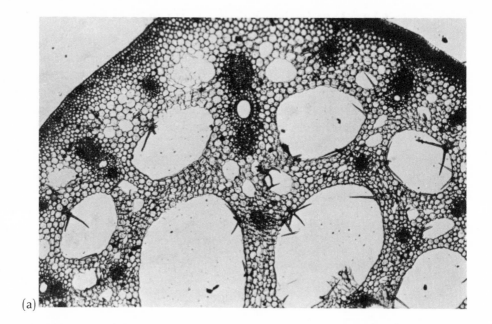

(a)

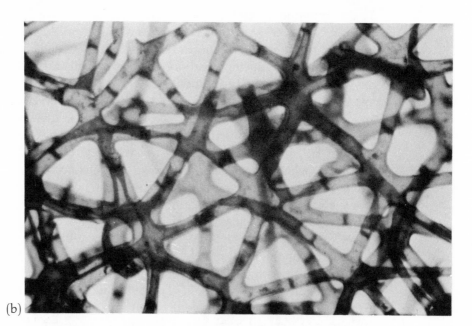

(b)

Figure 3.5.   (a) Transverse section of the submerged stem of *Nymphaea* (water lily). The large air spaces carry oxygen from exposed leaves to submerged parts of the plant.

(b) Stellate parenchyma of *Juncus* (rush) showing large air spaces between groups of cells.

3.2   Physical Properties of Soils

In so far as plant nutrition is concerned, the physical properties of solid phase particles are as important as their chemical state. Soils contain considerable quantities of inorganic ions which are bound to the charged surfaces of clay and organic soil particles. The gradual release of these reservoirs of bound ions provides a continuous supply of nutrients for plant absorption. In order to appreciate the nature of these dynamic exchanges it is necessary to explore the physical properties of soils, particularly adsorption and ion exchange. Soil particles, particularly the silicates have an overall negative charge and are capable of binding or adsorbing cations added to soil solution. Sand $(SiO_2)_n$ is composed of a three-dimensional array of silicate ions $(SiO_4)^{4-}$ in which each Si atom is surrounded tetrahedrally by four O atoms (Fig. 3.6). However, each O atom is covalently linked to a second Si atom so a giant neutral polymer results with an average formula represented as SiO_2. Imagine that some of the Si atoms in this structure be replaced by trivalent Al. The resulting aluminosilicate is negatively charged because some of the O valencies remain unsatisfied. For example, if 25% of the Si atoms are replaced the resulting structure would have the empirical formula $(AlSi_3O_8)^-$. With half of the Si atoms replaced the formula becomes $(Al_2Si_2O_8)^{2-}$. Mixed aluminosilicates of this kind are common among the primary minerals. These minerals may form open-framed structures into which

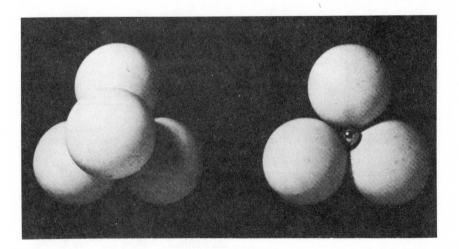

Figure 3.6.   Three dimensional representation of the SiO_2 polymer. Each S atom is surrounded by 4 oxygen atoms. (Photograph courtesy of the British Museum)

gases, liquids and small ions may diffuse. More importantly their negative charges may bind cations and participate in cation exchange reactions. Naturally, the smaller the particle size the greater is the relative surface area for these ion binding processes.

3.2a   ADSORPTION AND ION EXCHANGE

When a solution containing charged particles is brought into contact with a negatively charged surface the cations are attracted and bound to the surface whereas the anions are repelled. As a consequence the layer of solution close to the charged surface is enriched in cations and diminished in anions. Since the force of attraction or repulsion between charged particles diminishes with the square of the distances apart, the concentration of cations tends to decline exponentially with distance from the charged surface. The opposite situation applies to the distribution of anions. The distance over which these effects apply may vary from about 0.5 to 50 x 10^{-9} m. Figure 3.7 illustrates the

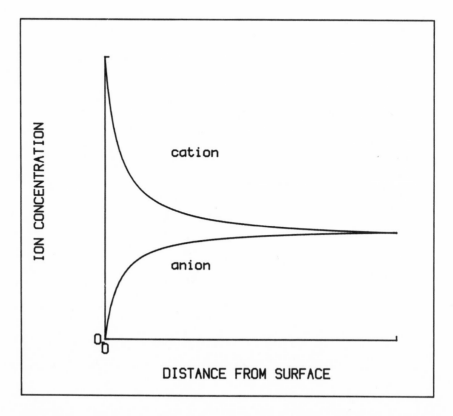

Figure 3.7.   Typical ion distribution adjacent to a negatively charged surface.

distribution of cations and anions in this layer and in the adjacent free solution.

Adsorption of ions to a charged surface is very much a dynamic process. Adsorbed ions with sufficient kinetic energy are able to overcome the electrostatic attractive force and leave the surface. Other ions take the place of those leaving the surface, so that at equilibrium equal numbers of ions move to and from the charged surface. Several factors influence the extent or tenacity of ion binding to a surface:-

(a) CONCENTRATION

As concentration in free solution increases, the adsorption increases until all potential binding sites are saturated. A plot of adsorption versus concentration takes the form of a rectangular hyperbola (Fig. 3.8).

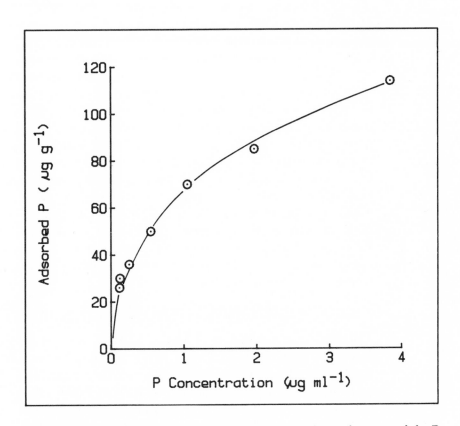

Figure 3.8. Typical pattern of P absorption to soil as a function of the P concentration of soil solution. (From Robertson, *Proceedings of the Alberta Soil Science Workshop*, 1980)

(b) CHARGE

Generally the tenacity of binding is in the order trivalent > divalent > monovalent ions. Nevertheless, there are exceptions to this rule. For example, in the mineral mica, K^+ is more strongly bound than Ca^{2+}.

(c) SIZE

Charge density of an ion decreases with increasing diameter. As a consequence small ions, with greater charge density, bind more water, creating larger hydration shells. The hydration layer tends to attenuate the forces of attraction between charged particles, and thus small particles tend to be more weakly bound to charged surfaces than larger particles.

The more strongly an ion is bound, the less likely it is to be replaced by a different ion species. The Hoffmeister cation series lists the binding capacity (or the tendency to displace other ions from adsorption) as $Ba^{2+} > Sr^{2+} > Ca^{2+} > Mg^{2+} > Cs^+ > Rb^+ > K^+ > Na^+ > Li^+$. Nevertheless, a weakly adsorbed ion may replace a more strongly bound ion if the former is present at sufficiently high concentration. This is possible because the kinetic energies of populations of a particular ion, e.g. K^+ or Ca^{2+}, tend to be quite variable. Most ions have energies close to the mean for the population, but a small proportion will have values much higher or much lower than the mean. Even though K^+ is generally unable to displace Ca^{2+} from adsorption on a 1:1 basis, increasing $K^+:Ca^{2+}$ ratio in free solution increases the probability that one of the K^+ ions will have sufficient kinetic energy to displace Ca^{2+}. Thus replacement of one ion species by another is a matter of statistics in addition to the other considerations we have discussed. Figure 3.9 shows how a cation exchange resin, initially in the acid form, can be used to remove K^+ from a KNO_3 solution. To regenerate the acid form of the column and displace the K^+ ions it is necessary simply to elute the column with a strongly acidic solution.

3.2b ION EXCHANGE PHENOMENA IN SOILS

The preceding discussion of ion adsorption and ion exchange is extremely relevant in so far as the availability of ions for plant absorption is concerned. In addition to the negatively charged mineral particles, organic matter contains numerous carboxyl and hydroxyl groups in which H^+ may be replaced by other cations. Furthermore, soil particles contain some anion-binding sites. Hydroxyl ions, e.g. as in $Al(OH)_3$, may be replaced by various nutrient anions.

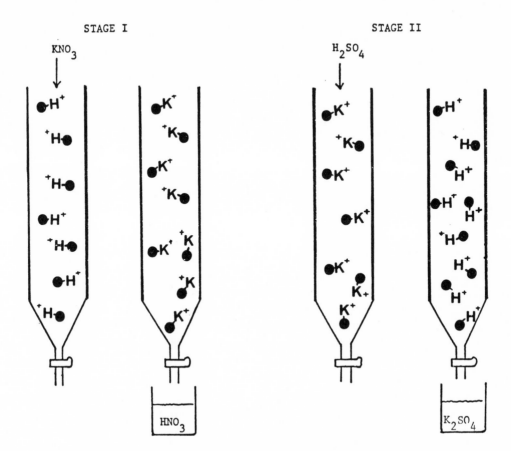

Figure 3.9.   Elution of an ion exchange resin

Stage I.   Ion exchange resin is in acid form. Addition of KNO_3 results in replacement of H^+ by K^+ and HNO_3 is eluted from column.

Stage II.   Addition of H_2SO_4 results in replacement of K^+ by H^+ and K_2SO_4 elutes from column. Resin is restored to acid form.

When fertilizers containing potash (K), lime ($CaCO_3$) and ammonium compounds are added to soil these cations enter into soil solution causing increased adsorption and even replacement of existing cations on the soil particles. So far as anion binding is concerned there is virtually no adsorption of monovalent anions such as NO_3^- and Cl^-, so that the NO_3^- reserve of the soil is limited to NO_3^- in soil solution and NO_3^- generated by nitrification. As a consequence NO_3^- levels in soil solution may fluctuate considerably. For the divalent anions, e.g. SO_4^{2-}, binding is a function of pH. Above pH 6.0-6.5 there is only very slight binding. In acid soils however, particularly those high in Fe and Al oxides, binding may be substantial. The binding or fixing of phosphorus is extremely important. This may involve precipitation by reaction with soluble Fe, Al or Mn salts:

$$Al^{3+} + H_2PO_4^- + 2H_2O \rightarrow 2H^+ + Al(OH)_2H_2PO_4 \text{ (hydroxy-}$$

phosphate) or replacement of hydroxyl groups:

$$\begin{array}{c} \quad\;\; OH \\ \quad\;\; / \\ Al - OH + H_2PO_4^- \rightarrow \\ \quad\;\; \backslash \\ \quad\;\; OH \end{array} \begin{array}{c} OH \\ / \\ Al - OH \quad + OH^- \\ \backslash \\ H_2PO_4 \end{array}$$

These bound or chemically fixed nutrients represent important reservoirs. Removal of nutrients from soil solution by root absorption tends to be compensated by release from these reservoirs:

Bound Nutrients → Free Nutrients → Absorbed Nutrients
(not directly available)   (in soil solution)   (by plant roots)

For example, soil solution contains only very low concentrations of phosphate ($\sim$ 1-10 μM). Yet rates of phosphate removal by plant roots (in a single day) may amount to several times the quantity present in soil solution. It has been calculated that the phosphate content of soil solution must be completely replaced as many as ten times per day under these conditions as the reactions in the above equation proceed toward the right hand side. Similar exchange processes are responsible for a continued provision of cations to soil solution.

A rather deleterious consequence of the ion exchange property of soils is to be seen in the effects of low pH. Some crop species, e.g. alfalfa, are unable to grow in acid soils (<pH 6), while oats and some barley varieties are, by comparison, relatively insensitive to such conditions. Yet in laboratory studies some "acid-sensitive" species tolerate low pH quite well when grown hydroponically. The explanation is to be found in the indirect effects of pH. In soils, bound Al^{3+} is displaced by H^+ at low pH resulting in Al-toxicity in susceptible plants. (This topic is discussed in greater detail in Chapter 9).

3.3 Nutrient Fluxes From Soil Solution to Root

3.3a THE GENERAL PHENOMENON

Nutrients in soil solution arrive at the surface of the root as a result of two kinds of movement; bulk flow and diffusion. According to the 2nd Law of Thermodynamics, spontaneous flows (fluxes) of matter or energy (i.e. those which occur without an input of free energy), can only occur where gradients of free energy exist throughout the system under study. These spontaneous fluxes occur along the gradient of free energy (from high to low free energy) until the free energy of the system becomes distributed homogeneously throughout. There can then be no further net movement within the system; it is at equilibrium. Consider the system depicted in Figure 3.10.

If A contains a gas which has a higher free energy (e.g. higher pressure) than in B, gas will flow from A to B, when the valve at C is opened, until the system comes to equilibrium and the free energy gradient from A to B is dissipated. Gas pressure will now be uniform throughout. Similarly, spontaneous fluxes of water, solute, heat, or

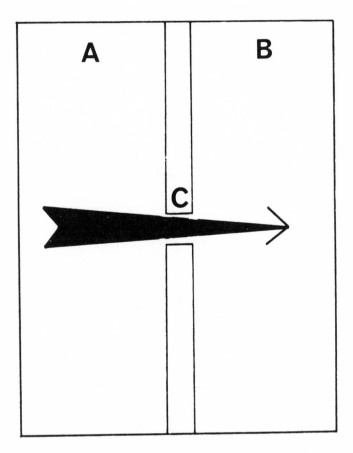

Figure 3.10. Hypothetical two compartment system containing a gas at higher pressure in section A than in section B. C represents a valve connecting A and B. Free energy and pressure gradients from A to B are shown by the arrow.

charged particles can only occur so long as these entities are non-uniformly distributed throughout the system. Thereafter no *net* movements can occur. In each case, the rates of these flows of matter or energy are proportional to the gradients of free energy (the driving forces) responsible for these fluxes. In the above examples the driving forces would be pressure gradient ($\frac{dP}{dx}$) water potential gradient ($\frac{d\psi}{dx}$), concentration gradient ($\frac{dC}{dx}$), temperature gradient ($\frac{dT}{dx}$) and electrical gradient ($\frac{dE}{dx}$). A general quantitative statement to describe these fluxes can be written as follows:

$$J \propto \frac{dZ}{dx} \tag{3.1}$$
$$\text{or } J = K \frac{dZ}{dx} \tag{3.2}$$

where J is the flux (e.g. mol m^{-2} s^{-1}), $\frac{dZ}{dx}$ is the gradient of the driving force responsible for the particular flux (i.e., the rate of change of the factor Z per unit distance), and K is the constant of proportionality linking the flux and the driving force (i.e., the flux per unit driving force). This simple equation will be familiar in the form known as Ohm's Law for electron flux:

$$I = \frac{E}{R} \tag{3.3}$$

where I = current, E = electrical potential difference and R = resistance (equivalent to 1/K of equation 3.2). Since the driving force for electron flux is the potential difference (E) this equation conform to the general statement above. It could be rewritten as:

$$I = C.E \tag{3.4}$$

where C = Conductance ($= \frac{1}{R}$). If the driving force were a gradient of concentration then we might rewrite the general equation as:

$$J = D \frac{dC}{dx} \tag{3.5}$$

D is analogous to K in Equation (3.2). If J is measured in mol m^{-2}s^{-1}, C in mol m^{-3}, and d in m, then D ($J \times \frac{dx}{dC}$) has the dimensions m^{2}s^{-1}. It is referred to as the diffusion coefficient. Equation (3.5) was first expressed in this form by Fick and is called Fick's first law of diffusion. As a matter of convention, since the flux occurs from high to low concentration, Fick's equation is usually represented as

$$J = -D \frac{dC}{dx}. \tag{3.6}$$

Where it is impossible to know the precise dimension of the diffusive pathway, as e.g. when a solute crosses a membrane we may write that

$$J = P \, dC \tag{3.7}$$

where P is the Permeability coefficient (units = m s^{-1})

In referring to the free energy of a solute we may use the term chemical potential (μ) which is the free energy mol^{-1} (Joules mol^{-1}). This quantity is unique for a particular solute under standard conditions but bears a direct relationship to concentration (actually activity) as follows:-

$$\mu = \mu^* + RT \ln C \qquad (3.8)$$

where μ^* is the standard state chemical potential, R is the gas constant (8.3 Joules mol^{-1} degree^{-1}), T is the absolute temperature in degrees Celsius and $\ln C$ is $\log_e$ of concentration. Thus, in the case of a diffusive flux of matter we may refer to the flux occurring along the chemical potential gradient ($\frac{d\mu}{dx}$).

One last example deserves consideration because it will crop up in relation to electron fluxes during reduction/oxidation (redox) reactions. Just as the chemical potential of a solute is a measure of its free energy content so the term redox potential (measured in units of volts) is a measure of the free energy (capacity for electron transfer) of a reducing agent. The more negative the redox potential, the greater is the free energy and the reducing capacity of the molecule. Thus ferredoxin, a reducing agent involved in the light reactions of photosynthesis has a redox potential of −0.43 volts. NADP$^+$, which is reduced by electrons received from reduced ferredoxin, has a redox potential of −0.32 volts.

3.3b THE SOIL SOLUTION CASE

Bulk flows of soil solution occur when the soil solution moves *en masse* from one region of the soil to another. For example, a rapidly transpiring plant draws a substantial bulk flow of soil solution up to the root surface. The rate of this flow of solution is given by the equation:

$$J_w = Lp \, \Delta \Psi \qquad (3.9)$$

where J_w = rate of bulk flow of solution, Lp (hydraulic conductivity) is a coefficient analogous to the conductance term in equation 3.2, and $\Delta \Psi$ = difference of water potential. If J_w is in m^{3}m^{-2}s^{-1} (= m s^{-1}) and $\Delta\Psi$ is in bar, then Lp (= $\frac{J_w}{\Delta\Psi}$) is measured in m s^{-1}bar^{-1}. Nutrients in solution will naturally be carried along by this flow. Quantitatively, the rate of delivery of an ion to the root by bulk flow is a function of the concentration of the ion in solution and the rate of bulk flow. This flux of ions (J_c) can be calculated as follows:

$$J_c = C \cdot J_w \qquad (3.10)$$

where J_c = bulk flux of ion

C = concentration of ion in solution

J_w = rate of bulk flow of solution. If J_c is small by comparison with the capacity for uptake by the root then bulk flux will be inadequate to satisfy the plant's needs. This may apply when J_w is low, as in dry soils, or when C is low as is generally the case for Pi. However, if plant demand is small by comparison to J_c then the ion may become concentrated at the root surface. This latter effect can sometimes be responsible for a build up of Ca^{2+} at the root surface.

Diffusion represents the second means whereby nutrients can move in soil solution. This is independent of any bulk flows. Diffusion of a dissolved solute occurs whenever a difference of chemical potential exists between two locations. This diffusive flux may be represented by Fick's law (Equation 3.6, above).

Because the bulk fluxes of Pi and K^+ are commonly inadequate to supply plant demands, particularly under conditions of reduced transpiration when J_w is low, the supply of these ions by diffusion assumes the greater significance. As a consequence of rapid absorption at the root surface, steep gradients of concentration develop between soil solution at the root surface and bulk soil solution. The gradients may be so steep that the zone immediately adjacent to the root is almost depleted of ions as shown in Figure 3.11. Barber (1968) visualized these depletion zones by growing corn roots in soil labeled with ^{86}Rb (a radioactive tracer for K). Autoradiographs revealed depletion zones around the root in which levels of radioactivity were very low (see Fig. 3.12).

Where plant density is high, as in pastures and in many natural habitats, adjacent depletion zones will inevitably overlap. Under such conditions it has been proposed that rates of Pi and K^+ absorption by plant roots are determined by their rates of diffusion across the depletion zone, rather than by the roots' potential for uptake. While this is entirely correct in considering a static root system there are also important biological strategies which have evolved to reduce this diffusion limitation. For example, the rate of diffusion across the depletion zone will be a function of the ion's concentration gradient and the diffusion coefficient. The more efficiently the ion is absorbed at the root surface, the steeper this gradient will be.

Another important factor is the continued growth of the main root and the proliferation of branch roots and root hairs which bring the plant into contact with 'unexplored' soil. Infection of the root with various (mycorrhizal) fungi provides an important mechanism for extension of the root's zone of absorption. As a consequence different species grown under identical conditions may absorb quite different

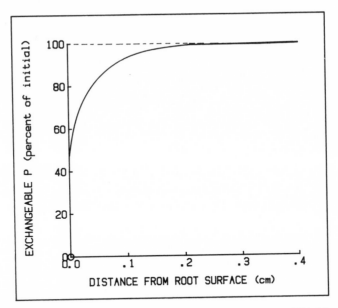

Figure 3.11. Concentration of exchangeable phosphorus (Pi) (shown as a percentage of that initially present) as a function of distance from the root surface (From Bhat and Nye, *Plant and Soil*, 41: 383-394, 1974)

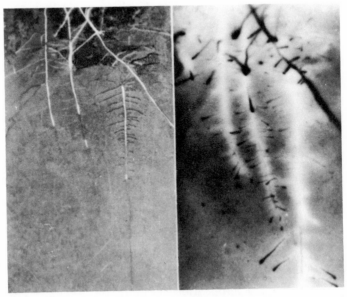

Figure 3.12. A photograph (left) and an autoradiograph (right) showing the effect of corn roots on the distribution of ^{86}Rb in the soil. Light areas show ^{86}Rb depletion around the corn roots. (From Barber, in *The Role of Potassium in Agriculture*, American Society of Agronomy, 1968)

quantities of nutrients from the same soil. These topics will be explored more fully in the next chapter.

3.4   Biological and Fertilizer Inputs to Soils

3.4a   NITROGEN FIXATION

In addition to the contributions from the original bedrock, substantial chemical inputs to a soil may be derived from various biological processes and through fertilizer applications by man. The fixation of atmospheric N_2 by free-living and symbiotic microorganisms is estimated to contribute approximately 2×10^{11} kg of N to soils annually. This is about four times greater than the figure for all industrial fixation of nitrogen! The organisms associated with legume root nodules (*Rhizobium* species) and with alder roots (*Actinomyces alni*) are particularly important in this context. The enzyme responsible for the reduction of N_2 to NH_3 is called nitrogenase. Reducing power and ATP for the reduction are provided by the oxidation of carbohydrate, which is supplied in symbiotic systems by the host plant. The overall reaction requires 12 moles of ATP and 6 electrons (supplied by reduced ferredoxin) per mole of N_2 reduced. Like the industrial counterpart (the Haber-Bosch process) this reaction is energetically expensive. The reaction sequence, composed of at least 3 steps, is represented in the overall equation below:

$$N_2 + 6H^+ + 6e^- \xrightarrow[\qquad]{\quad 12\,ATP \qquad 12(ADP + Pi) \quad} 2\,NH_3$$

The nitrogenase enzyme consists of 2 protein components. One, the Fe protein is made up of 2 identical subunits and 4 atoms each of Fe and S. The other component, the Mo-Fe protein consists of 4 subunits and contains Mo, Fe and S. It is thought that ATP is consumed during N_2 reduction to bring about reduction of the Fe protein. This in turn reduces the Fe-Mo protein which is responsible for the reduction of N_2.

3.4b   INORGANIC FERTILIZERS

Agriculture in industrially developed countries is heavily dependent upon the application of inorganic fertilizers. In the USA approximately 10 million tonnes of N, 5 million tonnes of P and 6 million tonnes of K fertilizers are used annually, at a cost of about 7 billion dollars. Figure 3.13 shows the increasing global consumption of N ($\square$),

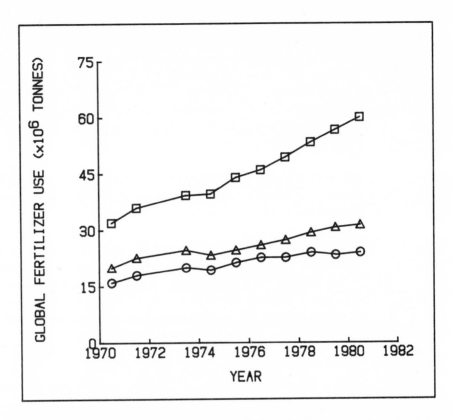

Figure 3.13. Global fertilizer use of N(□), P (△), and K (O) in millions of tonnes (1 tonne = 10^3 kg) for the period from 1971 to 1981. (*Source: F.A.O. Fertilizer Yearbook*, 1983)

P (*Δ*) and K (o) fertilizer during a 10 year period to 1981. Roughly one half of this use is accounted for by USA and Europe. In addition to improving yields, fertilizers can make infertile soils productive by remedying deficiencies of specific elements or by ameliorating extremes of pH. For example, in the Peace River region of Alberta where some soils have a pH of 4.5 the yield of Bonanza barley was increased from 35 to 2907 kg ha^{-1} through applications of lime at the rate of 18 tonnes ha^{-1}. This application raised average soil pH to 6.2.

3.5 The Aquatic Systems

Plants which inhabit marine and fresh water systems generally obtain nutrients directly from the aqueous environment and only a few species in shallower waters draw inorganic nutrients directly from the soil-like sediments. *Zostera marina* (eel grass), a monocotyledonous marine angiosperm (Figure 3.14) which occupies shallow, protected bays and lagoons, can absorb phosphate both through roots (anchored in the sandy sediment) and through the leaves. In these waters, as in ponds and shallow lakes, although nutrient exchange between sediment and overlying water is possible, the greater distances involved, compared to those in soil solution, make equilibration between these two phases extremely slow unless there is vigorous mixing of the waters as occurs in tidal action. In deeper waters there will be stratification of the water column because of light limitation of photosynthesis and temperature gradients. At the surface, where photosynthetic activity is vigorous, the rapid growth of algal species may reduce nutrient concentrations to much lower levels than are found in deeper waters. The removal of NO_3^- by phytoplankton may deplete the surface waters to such an extent by early summer that growth is essentially nitrogen-limited. Surface waters are also warmer and less dense than those at greater depth and hence mixing by convective movement is precluded. Winter storms cause vigorous mixing of the upper and lower layers of waters, restoring nitrogen levels so that a new flush of growth is possible in spring when temperatures begin to rise. Generally, concentrations of the limiting nutrients N and P are

Figure 3.14. *Zostera marina* (eel grass) photographed on the coast of British Columbia outside Vancouver. (Photograph courtesy of Dr. P.G. Harrison)

considerably lower in the aqueous environment than in soil solution. Table 3.1 gives representative values for concentrations of the major nutrients in soil and in the marine and fresh water ecosystems.

Summary

Soils consist of solid, liquid and gaseous phases. The solid phase generally contains the major inorganic reserves, while soil solution represents the immediate source of nutrients for plant absorption. The gaseous phase provides a source of oxygen for root activity.

The physical properties of soil, which are strongly influenced by surface area to volume relations of the soil particles, are extremely important in determining the availability of various ions to plant roots. Generally ions move to the root surface by bulk (convective) flows and by diffusion. When bulk flows are insufficient to supply plant demand, diffusion-limited zones may develop adjacent to the root. Root growth into unexplored regions of soil represents an important means of resolving this diffusion-limited situation.

Further Reading

Barber, S.A. 1974. Influence of the plant root on ion movement in soil. In *The Plant Root and its Environment*, ed. E.W. Carson. Charlottesville: University Press of Virginia.

Barber, S.A. 1984. *Soil Nutrient Bioavailability: a mechanistic approach*. New York: Wiley.

Brady, W.C. 1974. *The Nature and Properties of Soils*. 8th ed. New York: MacMillan.

Mengel, K., and Kirkby, E.A. 1979. *Principles of Plant Nutrition*. Berne: International Potash Institute.

Nye, P.H., and Tinker, P.B. 1977. *Solute Movement in the Soil-Root System*. Oxford: Blackwell Scientific Publications.

Scott Russell, R. 1977. *Plant Root Systems. Their function and interaction with soil*. London: McGraw-Hill.

Stolzy, L.H. 1974. Soil atmosphere. In *The Plant Root and its Environment*, ed. E.W. Carson. Charlottesville: University Press of Virginia.

4

Root Structure in Relation to Inorganic Nutrition

Roots perform four major functions in the lives of plants:

(1) They are responsible for the absorption of inorganic nutrients and water from soil solution,

(2) they bring about the radial transport of absorbed materials across the root cortex to the xylem, for axial transport (translocation) to stems and leaves,

(3) they anchor the above-ground parts in soil, and

(4) they serve to store inorganic and organic nutrients.

In addition, roots participate in numerous important biochemical processes. For example, significant quantities of NO_3^- and SO_4^{2-} are reduced in roots of some species. Synthesis of certain hormones and (in some species) alkaloids occurs within root tissues. Roots are also capable of serving as a barrier which protects the shoot from chemical damage. Potentially harmful chemicals such as phenolic compounds, salt (NaCl) and heavy metals, may be detoxified, either by being sequestered in particular compartments (e.g., vacuoles of cortical cells) or by chemical modification. In this chapter root structure is discussed in relation to inorganic nutrition. This will involve consideration of root morphology, anatomy and even the ultrastructure of particular cells.

4.1   Root Morphology

We have seen from Chapter 3 that movement of nutrients from soil to root may be so limited, by various soil factors, that depletion zones develop around the root and nutrient absorption in those regions may then depend upon rates of diffusion across these zones. However it is a misconception to believe that acquisition of nutrients by the plant is entirely determined by such abiotic factors. Through continuous growth into fresh 'unexplored' regions of soil this problem can be substantially alleviated. Moreover, the pattern of root growth, particularly the overall shape and density of rooting, can be extremely important in determining rates of nutrient acquisition.

When well spaced plants grow into moist, chemically and physically homogeneous soil, free from large stones and rocks, their root systems map out 3-dimensional shapes which may approximate hemispheres, cylinders or cones (Fig. 4.1). Typically the main root or adventitious roots will grow vertically into the soil. When lateral branches (first-order laterals) emerge these tend to grow out at right angles to the parent organ, often dipping vertically at some distance from the branch point. First-order ($1°$) laterals bear second order ($2°$) laterals and in turn the latter may bear third-order laterals. Through this extensive branching a large proportion of the soil volume may be brought into intimate contact with the root, facilitating continued absorption of nutrients.

Nevertheless, there is considerable variation in patterns of root morphology among different plant species, different habitats and in response to environmental variations (Fig. 4.1). Dicots typically generate tap root systems (see below) consisting of a relatively small number of large roots. By contrast, monocots tend to give rise to fibrous root systems, made up of large numbers of finely divided roots which may form a dense mat in superficial soil layers.

Within plant communities there may be stratification of root systems so that different species tend to draw water and nutrients from different layers, avoiding direct competition. Some plants may have roots located in both superficial soil layers and in layers several metres below. Alfalfa roots, for example, may penetrate to a depth of

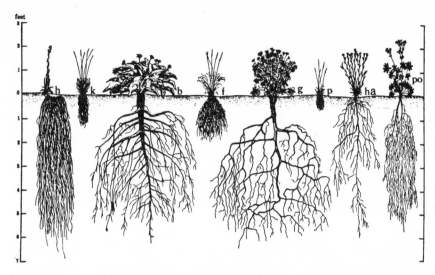

Figure 4.1. Schematic bisect showing root and stem relations of several prairie plants: h, *Hieracium scouleri* (hawkweed); k, *Koeleria cristata* (June grass); b, *Balsamorhiza sagittata* (balsam root); f, *Festuca idahoensis* (fescue); g, *Geranium viscosissimum* (geranium); p, *Poa sandbergii* (bluegrass); ha, *Haplopappus racemosus*; po, *Potentilla gracilis* (cinquefoil). (From Weaver, in *The Ecological Relations of Roots*, Carnegie Institution of Washington, 1919).

6 or more metres. Generally, superficial layers are richer in nutrients and there tends to be a greater root biomass and greater absorption from these layers. However, patterns of root development are strongly influenced by soil moisture, temperature, nutrient availability, and competing plant species. For example, as upper layers are depleted of nutrients roots tend to draw relatively greater proportions of nutrients from greater depths (see Fig. 4.2).

Plant adaptation to particular habitats has resulted in characteristic root morphologies. Thus, plants which occupy sandy soils commonly possess numerous lateral roots which spread horizontally, close to the soil surface. There is also considerable variation in root morphology which arises from localized environment%riation. Compacted regions of soil and stones and rocks present barriers to direct penetration by growing roots. An extremely adaptable root morphology, particularly responsive to local variation, enables the root to bypass such local impediments. Similarly, localized supplies of nutrients evoke extensive root proliferation into the enriched region of soil enabling the plant to obtain the maximum available resources. This morphological plasticity is in contrast to shoot morphology with its precise leaf arrangement (phyllotaxy).

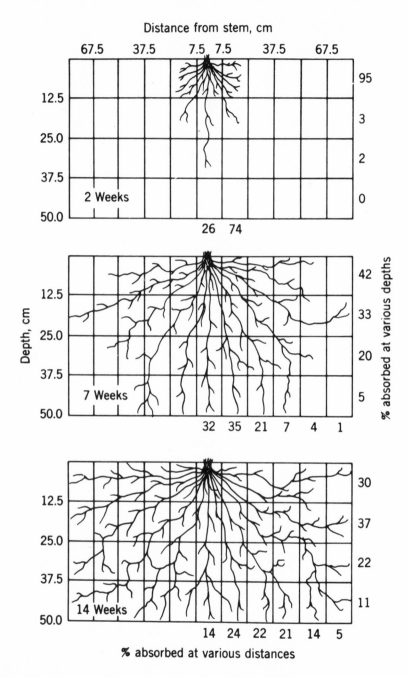

Figure 4.2. Diagrams showing the effect of extension of corn root systems on the amounts of ^{32}P absorbed at various soil depths and distances from the plant by root systems of various ages. (From Hall et al., North Carolina Agriculture Experimental Station Technical Bulletin 101, 1953)

4.2 Origin and Development of the Root System

The polarity of the embryonic root/shoot axis is established early on during embryogenesis and by the time that seed development is complete an embryo root (the radicle), which may possess branch roots, is already formed. This first root is generally called the primary root. During germination the elongation of the radicle causes it to break through the seed coat. Figure 4.3 shows the primary root of a

Figure 4.3. A root tip of a radish seedling showing the root cap and root hairs at various stages of development. (Glass, unpublished)

radish seedling with numerous root hairs. The continued elongation of this primary root, together with the emergence of numerous lateral roots, which themselves elongate and bear lateral branches, soon gives rise to an extensive root system. In dicots the entire root system is typically derived from the radicle in this manner, and even at maturity the primary root is large by comparison with 1°, 2° and 3° order laterals. Such an arrangement is referred to as a tap root system.

In monocots, by contrast, the primary root is short-lived, and soon after germination it is replaced by numerous adventitious roots which originate at the base of the stem. These roots, together with their numerous laterals, constitute fibrous root systems, in which no one root predominates over others (see Fig. 4.2). Some years ago Dittmer estimated the extent of root growth in a greenhouse-grown winter rye plant, grown for 4 weeks in 2 cubic feet of soil. The combined length of the main root plus 1° and 2° laterals was 385 miles, with a surface area of 2,554 square feet! Root hairs numbered 14.5×10^9, with total length equal to 6,603 miles and surface area of 4,321 square feet, providing a total below-ground absorptive surface which was roughly 130 times that of above-ground surfaces.

Unlike the shoot, where appendages, (leaves and axillary buds) are developed from primordia differentiated by the apical meristem, appendages of the root (except for root hairs) develop at some distance back from the apical meristem from cells within the stele.

The variability and morphological complexity of root systems should not be construed to imply that the root develops in a haphazard, uncontrolled manner. Rather, the complexity and variability must surely reflect the operation of an extremely delicate and adaptable morphogenetic system.

4.3 Root Anatomy

Roots extend in length through cell growth in the regions immediately behind the root tip. The root apex (Fig. 4.4) typically consists of a dome-shaped root cap which is usually less than 1 mm in length. The root cap is made up of parenchymatous cells which are generated by divisions of the apical meristem, located behind the cap. The cap is thought to protect the meristem as the root 'forces' its way through soil. The cells of the root cap are continuously being sloughed off during root extension and their decomposition, together with the

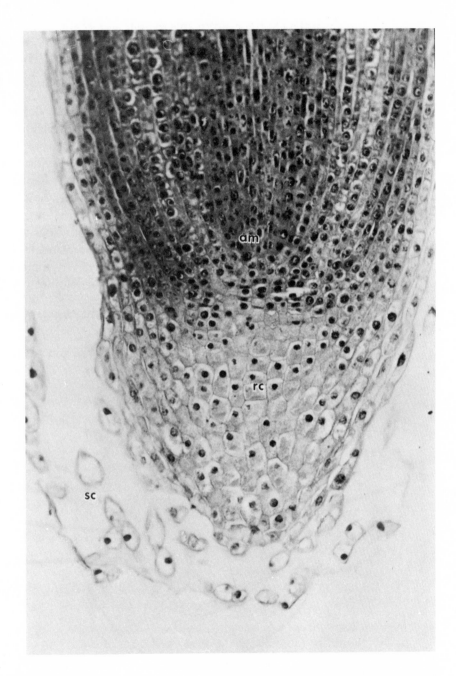

Figure 4.4   The root tip of a lentil seedling in longitudinal section (magnification X 240): sc, sloughed-off cells; rc, root cap; am apical meristem. (From Roland and Roland, *Atlas de biologie végétale*, Masson, 1980)

mucilage they secrete, lubricates the passage of the root through soil. This mucilage layer, sometimes referred to as mucigel, also provides nourishment for a rich flora of soil micro-organisms. The root cap is also extremely important as the organ responsible for perceiving gravity.

Behind the root cap the root apical meristem produces cells which differentiate into the epidermis, the cortex and the stele, all behind the meristem, as well as producing cells which form the root cap in front. It is important to appreciate that cell divisions *per se* do not produce extension of the axis. Rather, cells produced by the meristem constitute the raw material for extension and differentiation. Behind the meristem, in the region of elongation (1-10mm long), there is a massive increase of cell volume in the longitudinal axis. This is associated with a considerable uptake of water and inorganic ions and the formation of large central vacuoles. In mature cells the vacuole occupies 80-95% of cell volume, while cytoplasm is limited to a thin parietal layer, 1-2 μm in thickness.

The region of cell elongation merges into the region of maturation (1 to several cm long). Here continued differentiation gives rise to three concentric layers, namely epidermis, cortex and stele (Fig. 4.5a).

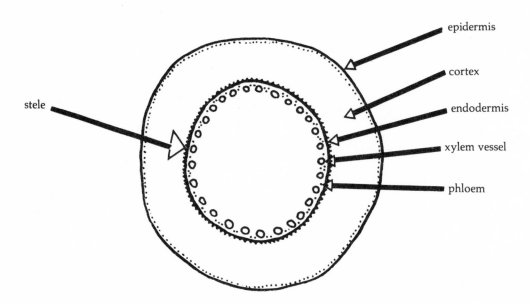

Figure 4.5 (a).   Transverse section of a corn root under low power magnification. (Drawings by L.R. Bohm)

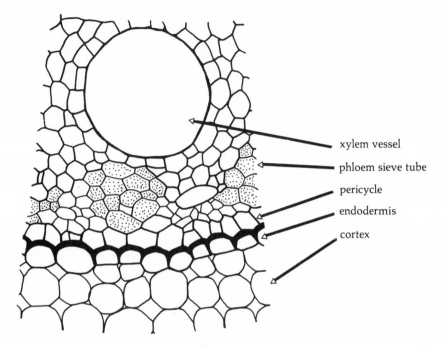

Figure 4.5 (b).   Magnified sector of corn root (high power magnification) to show vascular tissue, endodermis and cortical cells.

THE EPIDERMIS

The epidermis, the outermost cell layer, is particularly important for nutrient absorption as it gives rise to root hairs (Fig. 4.3). Each root hair is approximately 0.1 to 1 mm in length and there may be as many as 10^3 root hairs per cm of root. Generally root hairs emerge in the zone of maturation and and may still be found several cm back from this region. Intuitively it might seem obvious that root hairs are involved in nutrient uptake, however some workers have stressed that roots can absorb nutrients quite as efficiently without hairs. Bole (1973), for example, found that varieties of wheat with few root hairs absorbed phosphate as efficiently as varieties with many root hairs. Nevertheless, other workers have suggested that root hairs may enhance K^+ uptake by as much as 78% when this ion is present at low concentrations. Here, as in other cases in plant nutrition, we must take care to consider the conditions under which experimental findings are obtained, and be aware that laboratory studies usually greatly oversimplify the soil situation. Where nutrients are present at low concentration root hairs considerably extend the volume of soil which can be drawn upon for nutrient absorption. However, when nutrient concentrations are high depletion zones are considerably diminished and root hairs may be relatively less important. In the case of phosphate absorption, root hairs have been suggested to increase uptake by 2 to 3 fold even at high P levels. Using soil labeled with radioactive phosphate, depletion zones for this nutrient were found to be considerably enlarged by the presence of root hairs.

Environmental factors can exert a strong influence upon root hair development. For example aquatic plants and plants grown hydroponically generally have few root hairs. Above the water level, however, hydroponic plants may develop abundant root hairs. When hydroponically-grown barley plants are starved of phosphate for some time extensive root hairs develop even below the surface of the nutrient solution. It is tempting to suggest that where depletion zones are minimized by adequate mixing (in hydroponic systems) root hairs are suppressed. However, it should not be overlooked that under natural conditions root hairs also serve in absorbing water and providing anchorage. Moreover, in legumes the root hair is the site of infection by the *Rhizobium* responsible for nodule formation. In groundnut (a legume) it has recently been demonstrated that non-nodulating lines fail to produce root hairs.

Another important anatomical development of epidermal cells are the extensive ingrowths of the cell wall which sometimes occur in response to Fe-deficiency, salt stress or even mycorrhizal infection (see Section 4.5). Cells with such wall proliferations (referred to as transfer cells) were first identified within the stele, associated with conducting tissues (sieve tubes or xylem vessels). Figure 4.6 shows transfer cells adjacent to xylem vessels of a dicot root. The increased area of the wall and the plasma membrane (which is closely associated with the wall protuberances) together with numerous intercellular cytoplasmic connections (plasmodesmata) that traverse the wall, linking adjacent cells, would seem to be an ideal device for increasing transport between cells.

By the use of electron microscopy, Leppard has demonstrated the presence of numerous rhizoplane fibrils at the root surface in wheat roots. These fibrils, consisting of pectic substances, may be aggregated into tufts which are thought to provide intimate association with soil particles (see Fig. 4.7). These hitherto unknown structures facilitate cation exchange between soil particles and the root.

wall ingrowths   xylem parenchyma cell   plasma membrane

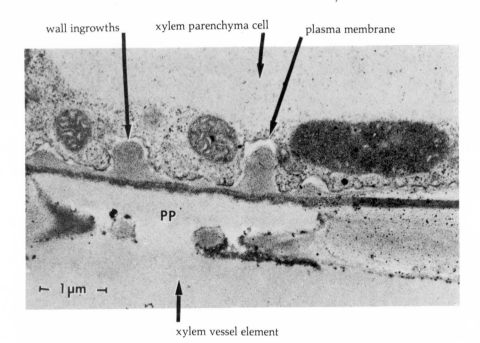

PP

⊢ 1 μm ⊣

xylem vessel element

Figure 4.6. Wall ingrowths of a xylem parenchyma (transfer) cell of soybean root opposite the primary pit area (PP) of an adjacent xylem vessel. Note how the plasma membrane follows the contours of the wall ingrowths. P, proplastid. (From Läuchli *et al.*, *Membrane Transport in Plants*, 1974)

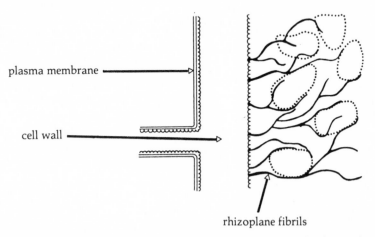

plasma membrane

cell wall

rhizoplane fibrils

Figure 4.7. Diagrammatic representation of the physical relationships at the root surface between root cells, cell walls, a tuft of rhizoplane fibrils and soil particles of the rhizosphere. (From Leppard and Ramomoorthy, *Canadian Journal of Botany*, 53:1729-1735, 1975)

THE CORTEX

Cells of the cortex make up a large part of the root volume in young roots (see Fig. 4.5). These cells are parenchymatous and the tissue possesses numerous intercellular spaces. The cytoplasm of adjacent cortical cells is linked by means of plasmodesmata, (see Figs 6.3 and 6.4) forming a cytoplasmic continuum called the symplasm that traverses the cortex, from epidermis through to the stele. The endodermis constitutes the innermost layer of the cortex. The radial and tangential walls of this layer are unique in possessing a water and solute impermeable band, composed of suberin and lignin, which completely encircles the cell wall. This structure is called the Casparian band or strip. Figure 4.8 provides a view into an endodermal cell as it

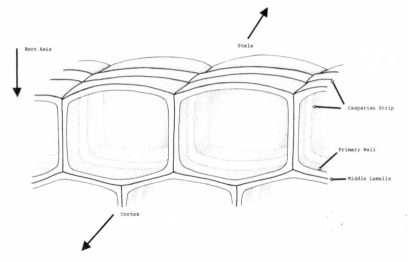

Figure 4.8(a).   State I endodermal cells in longitudinal tangential section. The diagram provides a perspective looking into an endodermal cell from the cortex. (Drawings by L.R. Bohm)

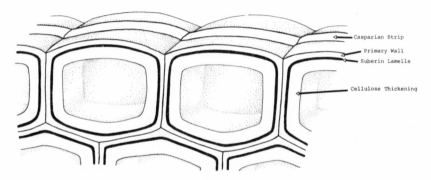

Figure 4.8(b).   State III endodermal cells in longitudinal tangential section.

would appear if we were able to look toward the stele from the cortex (a longitudinal tangential section). The plasma membrane of endodermal cells appears to be strongly bound to the cell wall in the region of the Casparian strip because even following plasmolysis this connection remains intact. This link between the membrane and the Casparian strip is demonstrated in the transverse section of an endodermal radial wall (Fig. 4.9). Adjacent endodermal cells are also tightly adjoined, without intercellular spaces. As a consequence of

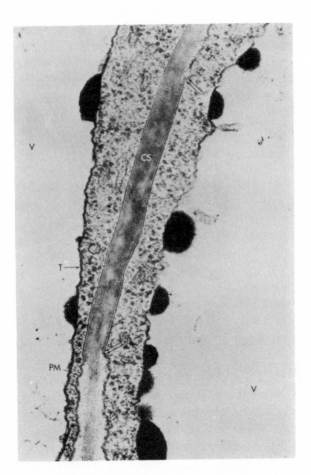

Figure 4.9.   Transverse section of the radial walls separating two endodermal cells of *Limonium* root (magnified X 140,000). Note the Casparian strip (cs) and its association with the plasma membrane (PM) shown by arrows at the centre of the micrograph. Lower down the wall is typically composed of a central middle lamella and adjacent primary walls. T, tonoplast; V, vacuole. (From Ledbetter and Porter, in *An Atlas of Plant Structure*. Springer-Verlag, 1970)

these features, water and solutes are largely unable to pass from the cortex to the stele without entering the endodermal cytoplasm. It is commonly overlooked that movements in the opposite direction (stele to cortex) are similarly controlled by these structures. Endodermal cells in which wall thickening is limited to the Casparian band are referred to as State I cells (as in Fig. 4.8(a)).

In many dicot roots, where secondary growth occurs within the stele, endodermal cells remain in this State I condition and are eventually lost as the central cylinder of secondary tissue expands, crushing the cortex. Where the primary condition is retained (as in monocots), a suberin layer is layed down on the entire inner surface of the endodermal primary wall, converting endodermal cells to the State II condition. Within 5 cm of the root apex of barley 50% of endodermal cells may be in this State II condition. Inside the suberin lamella is deposited a cellulose layer which may become heavily lignified. Endodermal wall thickenings are sometimes deposited uniformly on all inner surfaces (see Figs. 4.8(b) and 6.3). In other cases the thickenings are more extensive in the radial and inner tangential walls so that a characteristic 'C'-shape is produced in cross section. This cellulosic thickening brings endodermal cells to their State III condition.

In some cases the development of the additional wall thickenings of endodermal cells appears to be delayed, particularly in regions opposite xylem bundles, so that thin walled "passage cells" are found adjacent to heavily thickened endodermal cells. Their appearance and location, adjacent to xylem vessels has led anatomists to suggest that passage cells may represent low-resistance sites for the transfer of water and nutrients between the cortex and the stele. By contrast, Clarkson et al. (1971) concluded that in barley roots these cells made only a minor contribution to water and Pi entry into the stele.

THE STELE

Within the endodermis the first layer of thin walled parenchyma constitutes the pericycle. This layer retains meristematic potential and at regions adjacent to conducting tissues, (xylem and phloem), the pericycle may become active, giving rise to lateral root primordia. The growth of these primordia produces lateral roots which literally grow through the cortex and epidermis. These roots continue to extend outside the parent root and soon bear root hairs and subsequently lateral roots. Within the pericycle the conducting tissues, xylem and phloem, are located as alternating bundles around the periphery of the stele in the young root. In dicots there are 2 to 5 groups of xylem and phloem, whereas monocots tend to have many more (Fig. 4.5). As the dicot root ages a lateral cambium develops between xylem and phloem which is responsible for producing secondary xylem and phloem. In monocots, by contrast, no such development occurs.

Xylem vessels and tracheids are heavily lignified. At maturity these cells are non-living and serve as a passageway for the conduction of water and inorganic nutrients to the stem. The parenchymatous cells which are found adjacent to xylem vessels are thought to be important in transferring inorganic nutrients to the xylem conducting elements. This topic will be dealt with in more detail in Chapter 6. However it should be stressed that these cells may possess wall proliferations (typical of transfer cells) which may assist in their proposed function of delivering nutrients to the xylem.

4.4 Nutrient Absorption by the Root

4.4a LOCATION OF ABSORPTION ZONES IN THE LONGITUDINAL AXIS

It has frequently been stated that the root tip or the region of root hair development constitutes the principal zone for nutrient absorption. Much confusion has been generated by failing to distingush between absorption, accumulation and translocation. The term absorption (or uptake) refers to the amount of solute which is removed from the external medium. Translocation refers to the quantity of absorbed solute which is transported from the root, along the stele to the stem. By contrast, accumulation represents that part of absorbed solute which is not translocated but is retained within the root.

Experiments with numerous plant species have shown that absorption of nutrients can occur effectively for considerable distances behind the tip. For example, Clarkson (1974) has reported that rates of uptake of P and K were essentially constant over the entire length (>50 cm) of the roots of leeks. However we must be careful to distinguish between the potential for absorption (measured in the laboratory in solution-grown roots) and the actual absorption under natural conditions (in soil). Although the older (more basal) region of the root may be *capable* of absorbing nutrients, the removal of nutrients from soil in the vicinity of the root may so reduce soil solution concentration that there is little available for absorption. This issue is well illustrated in Figure 4.2 which shows the distribution of corn roots in soil as a function of age, together with data on the percent of ^{32}P which was absorbed from each region of soil at different intervals of time. As the plant aged more and more of its P was obtained at increasing distances from the stem, i.e. by the younger parts of the root.

4.4b LOCATION OF ABSORPTION ZONES IN THE RADIAL AXIS

An important question which may be posed at this juncture is "which cells of the root (from epidermis to endodermis) represent the major sites of nutrient absorption?" Many authorities favour the idea that the epidermis and all cortical cells up to the endodermis participate in absorbing ions from ambient solutions. This belief is based upon the notion that ions in solution penetrate the extracellular space from epidermis to endodermis very rapidly. This extracellular continuum is referred to as the apoplasm. There is certainly evidence in support of this, based upon experiments with hydroponically-grown roots and radioactive tracers for the nutrient ions. According to this perception, ions are absorbed from the apoplasm into the symplasm by cells all across the cortex. This would provide an extremely large total surface area for absorption. By contrast, other workers have proposed that nutrient absorption, particularly at the low concentrations which prevail under natural conditions, is limited to the epidermis. It has been argued that ions are so rapidly absorbed by the outer layers of the root (particularly the epidermis) that apoplasmic concentrations fall practically to zero by the time that the first or second cortical cell is reached. As a consequence, it is envisaged that ions enter into the symplasm only at the epidermis and traverse the cortex and endodermis by moving from cell to cell (within the symplasm) via plasmodesmata. These two possible routes across the cortex are represented diagramatically in Figure 4.10. Nevertheless, as external ion concentrations are raised to values approaching 0.2 to 0.5 mM, saturation of the absorption mechanisms (see Chapter 5) may permit ions to penetrate (within the apoplasm) to deeper cortical layers. Thus, the answer to this question depends very much upon prevailing conditions. My personal bias is toward the view that, in soil, nutrients would probably be most commonly absorbed by the outermost layer of the root. You may recall that depletion zones for K and P can develop around the root in soil so that even the epidermis may not receive adequate nutrient concentrations.

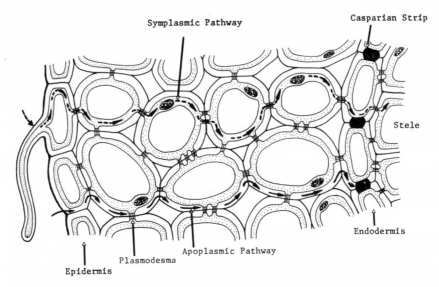

Symplasmic Pathway

Casparian Strip

Stele

Endodermis

Apoplasmic Pathway

Plasmodesma

Epidermis

Figure 4.10. A diagrammatic representation of the apoplasmic (solid lines) and symplasmic (dotted lines) pathways across the cortex of a typical root. Plasmodesmata are represented by shaded areas traversing the cell walls of adjacent cells. (Drawing by L. R. Bohm)

4.5 Morphological Adaptations in Response to Nutrient Supply

It is well documented that under conditions of nutrient deficiency root-to-shoot biomass ratio increases. Although the adaptive value of such a response is fairly obvious, the mechanisms responsible for this altered pattern of plant growth is not clear. When nutrients are in adequate supply, there is greater allocation of dry matter to the shoot than the root, with the exception of desert plants which may have extremely high root-shoot ratios and plants such as *Beta vulgaris* which have been specifically selected for root size. In barley e.g. the root:shoot ratio is normally about 0.25 and in tree species e.g. *Pinus* it may be as low as 0.19. However deficiencies of N, K or P as well as water stress cause this value to increase. In barley we have observed the ratio to increase from 0.25 to 1.0 as available K concentration was reduced.

It is now well known that there is differential root growth into regions of soil which contain localized concentrations of nutrients. The growth of the main root axis is little affected by nutrient deficiency but the initiation of lateral branches and their elongation rates may be substantially reduced. Drew and his associates have undertaken detailed studies of the modification of root form arising from localized supplies of N, P and other nutrients. Figure 4.11 shows the proliferation of lateral roots of barley into localized zones enriched in N and P. This kind of observation has considerable relevance for the placement of fertilizer under field conditions. Drew has reported that there was not only an increase of root biomass in the region supplied with nutrients (as in Fig. 4.10) but the rate of uptake of ^{32}P in this region was doubled on a per unit weight basis. It is evident, therefore, that location of nutrient supply can influence both physiological and morphological responses.

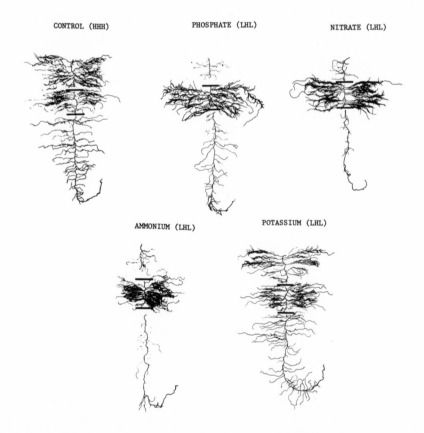

Figure 4.11.   Effect of a localized supply of phosphate, nitrate, ammonium, and potassium on root form in barley. Control plants (HHH) received a complete nutrient solution. The other roots received a complete solution only in the middle zone, top and bottom zones being supplied with a solution deficient in the specified nutrient. (From Drew, *New Phytologist*, 75:479-490, 1975)

4.5a MYCORRHIZAS

Under field conditions the roots of most plants are capable of forming symbiotic associations, termed mycorrhizas (literally fungus-roots), with soil fungi. These associations are particularly beneficial to the host plant, producing increased growth and yield under conditions of limited nutrient availability.

There are many different kinds of root/fungus associations but the most common and probably the most important are the vesicular-arbuscular (V.A.) mycorrhizas. These can be found in a large number of species including agricultural and horticultural plants. The fungi which produce V.A. mycorrhizas penetrate through the epidermis and cortex as far as the endodermis. The hyphae grow into individual cells and form the characteristic arbuscules (multi-branched hyphal structures) which are thought to participate in nutrient exchange between fungus and host plant. The fungi which form these associations may be obligate symbionts because (up to the present) it has been impossible to culture them on nutrient media.

Nevertheless, individual spores of these fungi can be isolated from soil and used to inoculate sterilized soil. In one experiment spores were isolated from various soils and used to coat seeds of onion and tomato, using methyl cellulose as an adhesive substance. Seeds treated in this way showed 30 to 40% yield increase compared to non-coated seeds when sown in sterilized soil (Table 4.1). Increased growth and yield, generated by this type of fungal association, has been demonstrated in both crop species and wild plants. The major emphasis has been upon the increased absorption of P by mycorrhizal roots because this element is commonly in limited supply in soils. However, it is also

Table 4.1. Influence of mycorrhizal fungi upon growth of onion and tomato plants in sterile soil to which mycorrhizal spores were added. Spores were pelleted on to the seeds of onion and tomato prior to planting. (From Gaunt, *New Zealand Journal of Botany*, 16:69-71, 1978)

	Dry Weights (g) per plant after 7 weeks	
Treatment	Onions	Tomatoes
Control (no spores added)	1.667	1.558
Seeds coated with spores	2.196	2.284
Percent increase due to spores	32%	47%

apparent that the absorption of other nutrients may be improved following "infection" by these fungi.

Studies with sterilized soils under field conditions and in pot trials in greenhouse experiments reveal that although adequate P may be present in the soil, the plant may be unable to acquire this nutrient in sufficient quantities to maintain potential growth rates without the fungal association (see Fig. 4.12). Nonetheless, when soil nutrient levels are raised significantly by high fertilizer application, growth may be unaffected by the presence of mycorrhizal fungi. Clearly, under such conditions the plant is able to obtain adequate amounts of

CONTROL   INOC.

Figure 4.12. Strawberry plants grown in low phosporous soil without (CONTROL) and with (INOC.) mycorrhizal fungi. (From Hayman, *The Plantsman*, 2:214-224, 1981)

P without the assistance of the fungus. In fact there is evidence that the plant/fungus association fails to develop under these conditions. Some researchers have proposed that under conditions of low nutrient availability the leakage of solutes (particularly sugars) from roots of the potential host plant provide the appropriate environment to nourish the fungus, encouraging the development of the mycorrhizal condition.

Several explanations for the increased nutrient absorption by mycorrhizal plants have been forthcoming. The hyphae of mycorrhizal fungi extend some considerable distance from the root into the surrounding soil. Using ^{32}P labeled soil it has been demonstrated that P up to 2cm away from the root was made available to the plant by the fungus. In addition the fungal hyphae present a considerably increased surface area for absorption and appear to have a higher affinity for P in soil solution than does the root itself. Calculations of the velocity of Pi transport within the hyphae (~ 2 cm h^{-1}) reveal that its progress is many times faster than diffusion rates in soil. Hence, considering the importance of diffusion-limited zones which normally develop in soil, this increased supply rate ultimately results in considerable increases in total P acquisition by the host plant. Where growth is P-limited this increased absorption results in increased growth rates. There is currently considerable interest in devising methods to screen mycorrhizal fungi for greater efficiency in nutrient uptake so that specific strains might be inoculated into soils to optimize resource utilization.

Summary

The acquisition of inorganic nutrients and water from soil represents one of the principal functions of roots. In mature plants, root morphology is extremely variable according to habitat and in response to local soil heterogeneities. Generally, through extensive branching, root hairs and mycorrhizal associations, the surface area of the root/soil interface assumes enormous dimensions. Rates of ion absorption in soil are probably highest in the younger, extending roots where depletion zones have not yet developed. Depending upon ion concentrations in soil solution, and plant demand for particular ions, absorption may occur at the epidermis followed by symplasmic transfer to the stele, or via apoplasmic transfer across the cortex and transfer through the endodermal symplasm to the stele. Regardless of which path is taken to the stele, ions (with minor exceptions) pass via the endodermal symplasm, the Casparian strip representing a permeability barrier to free movement to and from the stele.

Further Reading

Bracegirdle, B., and Miles, P.H. 1977. *An Atlas of Plant Structure*. London: Heinemann Educational Books.

Clarkson, D.T. and Robards, A.W. 1975. The endodermis, its structural development and physiological role. In *The Development and Function of Roots*, ed. J.G. Torrey and D.T. Clarkson. pp. 415-63. New York: Academic Press.

Esau, K. 1977. *Anatomy of Seed Plants*. 2nd ed. New York: Wiley.

Gunning, B.E.S., and Robards, A.W., eds. 1976. *Intercellular Communication in Plants: studies on plasmodesmata*. Berlin: Springer Verlag.

Kramer, P. 1969. *Plant and Soil Water Relations: a modern synthesis*. New York: McGraw Hill.

Läuchli, A. 1976. Apoplasmic transport in tissues. In *Encyclopedia of Plant Physiology, New Series, II B*, eds. U. Lüttge and M.G. Pitman, pp. 3-34. Berlin: Springer Verlag.

Roland, J-C., and Roland, F. 1980. *Atlas of Flowering Plant Structure*. London: Longmans.

Spanswick, R.M. 1976. Symplasmic Transport in tissues. In *Encyclopedia of Plant Physiology, New Series, II B*, eds. U. Lüttge and M.G. Pitman, pp. 35-53. Berlin: Springer Verlag.

Scott Russell, R. 1977. *Plant Root Systems. Their function and interaction with the soil*. London: McGraw Hill.

5

Ion Absorption

In this chapter we consider the processes involved in the transfer of inorganic ions from the external medium to the cell's interior. This involves passage through the cell wall and the major membranes of the cell, the plasma membrane and tonoplast. In arriving at current perceptions of these processes plant scientists have tended to concentrate their attentions upon a limited number of plant systems which have served as models for plant transport processes in general. For instance, algae such as *Nitella* and *Chara*, which possess internodal cells several centimetres in length, have been popular tools for study since the 1920's. Such exceptionally large cells are ideal for the direct sampling of subcellular compartments and for delicate microsurgical and microelectrical techniques. Among higher plants, seedling roots of cereals, particularly barley, and slices of carrot and beet root are widely used. These tissues are readily available, conveniently prepared for experimentation and generate extremely reproducible data.

In studying ion transport, plant physiologists have tended to focus attention upon the monovalent ions, notably K^+ and Cl^-. These have been favoured, not only because of their biological importance but also, because of the ease with which their concentrations can be measured. Moreover, unlike SO_4^{2-}, NO_3^- and Pi, which are metabolized to numerous organic derivatives following absorption, the monovalent ions retain their chemical identity, making the assessment of chemical potential gradients between various compartments considerably more straight-forward. The study of ion transport has been greatly facilitated by the availability of radioactive tracers which allow the researcher to measure rates of ion transport with great sensitivity. For example, using ^{42}K, a radioactive isotope of K, it is possible to measure fluxes as small as 10^{-7} mol K^+ h^{-1} g^{-1}.f.w. of root tissue. This is equivalent to 3.9×10^{-6} g of K^+! Radioisotopes such as ^{36}Cl, ^{32}P and ^{35}S are also widely used in ion transport studies. By contrast a small group of tracers, e.g. ^{13}N (half-life = 10 min), have such short half-lives (the half-life is the time required for half of the isotope to decay) that they can only be used in laboratories housed at the cyclotrons where these isotopes are prepared. Not surprisingly, such isotopes have seen only limited use.

5.1   Accumulation of Inorganic Ions

The selective accumulation of particular elements to levels which are many times in excess of their environmental concentrations represents one of the most characteristic features of plant cells. Figure 5.1 shows vacuolar concentrations of the major inorganic ions in *Nitella* together

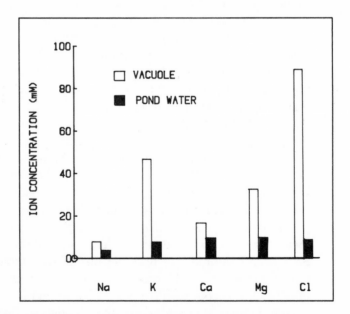

Figure 5.1.   Vacuolar concentrations of the major inorganic ions in cells of *Nitella* and in pond water in which this plant grows. (After Hoagland in Lectures on the inorganic nutrition of plants, *Chronica Botanica*, 1944).

with the concentrations of these same ions in the pond water supporting this plant.

In recent years several research groups have developed automated
systems for the maintenance of inorganic culture solutions at constant
composition. One such system is shown in Figure 5.2. Every 30 min
the computer opens solenoid valves (in sequence) that permit cali-
brating solutions to flow past ion-specific electrodes. Like pH (H^+)
electrodes, these can measure the concentration (activity) of ions such
as K^+, NO_3^- and Ca^{2+}. Electrode voltages, which are activity-
dependent, are measured by means of the DATA ACQUISITION
UNIT and used by the computer to generate calibrations curves for
each electrode. These curves are used to calculate ion concentrations
in the hydroponic tank(s) and to calculate what quantity of "top-up"
solutions must be added to restore ion concentrations to required
levels. In addition the computer keeps detailed records so that we can
monitor patterns of absorption (day and night) for several weeks.
Systems of this sort have enabled researchers to demonstrate that ions
such as K^+, NO_3^- and Pi can be accumulated from micromolar
concentrations in the medium to millimolar concentrations within

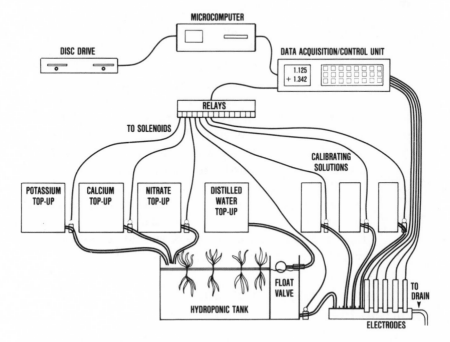

Figure 5.2. Automated computer-controlled system for measuring rates of
ion absorption by hydroponic plants and maintaining ion concentrations at
particular levels. (Glass, unpublished)

Table 5.1.   Potassium concentrations of tops and roots of 5 pasture species grown for several weeks in solutions maintained at 1, 8, 24, 95 and 1000 μm K^+. (From Asher and Ozanne, *Soil Science*, 103:155-161, 1967)

| Plant Species | K Concentrations ($\mu mol\ g^{-1}f.w.$) | | | | | | | | | |
| | Tops | | | | | Roots | | | | |
	1	8	24	95	1000	1	8	24	95	1000
Silver grass (*Vulpia myuros*)	32	150	<u>159</u>	174	205	16	52	<u>74</u>	85	103
Brome grass (*Bromus rigidus*)	0	156	<u>203</u>	226	245	13	48	<u>74</u>	88	132
Vetch (*Vicia sativa*)	22	101	<u>112</u>	121	142	12	29	<u>55</u>	53	86
Barrel medic (*Medicago tribuloides*)	40	106	<u>141</u>	155	202	9	23	<u>48</u>	56	104
Capeweed (*Cryptostemma calendula*)	16	8	101	<u>128</u>	164	9	70	77	<u>99</u>	143

Values underlined indicate treatments above which increases in the external potassium concentration gave no further increases in yield.

plant tissues. Table 5.1 provides data for the potassium concentrations of tops and roots of 5 pasture species grown in solutions maintained at 1, 8, 24, 95 and 1000 μM K^+ respectively. If we presume that most of the K^+ in these tissues is contained within the cells (i.e. apoplasmic content is small compared to that of the symplasm) and that cell volume is approximately 0.8 $cm^3\ g^{-1}$.f.w, then, for example, in silver grass tops the K^+ concentration is roughly 40 mM when grown at 1 μM K^+; this amounts to a 40,000 fold concentration! Similar findings have been documented for the other essential elements, particularly for P, N, S.

This extraordinary capacity to maintain a defined non-equilibrium composition of elements within plant tissues extends to the exclusion or extrusion of certain non-essential or toxic elements. Thus Na^+, Ca^{2+} or H^+ concentrations are often held at levels which are orders of magnitude lower than environmental distributions.

5.2   Fluxes in the Apoplasm: Apparent Free Space

Inorganic nutrients enter the apoplasm from soil solution either by diffusion along gradients of concentration or as a result of convective fluxes associated with the transpiration stream (Section 3.5). Depending upon prevailing conditions ions may penetrate the apoplasm as far as the endodermis (Section 4.4b). Moreover, the penetration of this cell wall space can be quite rapid. For example, when roots which had been briefly treated with ^{86}Rb (a tracer for K) were fixed and autoradiographed it was apparent that the isotope had equilibrated across the cortical apoplasm within a few minutes. Transport into the apoplasm (as distinct from transport across the plasma membrane, into the symplasm) can also be revealed kinetically. When roots or tissue slices are immersed in a $CaSO_4$ solution containing radioactive tracers, e.g. ^{42}KCl, there is initially a rapid movement of radioactivity into the tissue, as shown in Figure 5.3. Within minutes this declines to a much slower rate which can be sustained for several hours. Interestingly, if the experiment is repeated using roots pre-treated with a metabolic poison such as KCN, the initial rapid phase is retained but the subsequent slower phase is eliminated (Fig. 5.3). This rapid phase

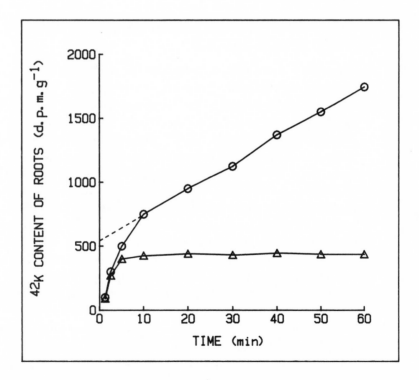

Figure 5.3. Penetration of ^{42}K into roots of barley plants immersed in a solution containing $100\mu M$ ^{42}KCl and 500 μM $CaSO_4$. O—control roots pretreated for 1 h with KCl; △—poisoned roots pretreated for 1 h with KCN. (Glass, unpublished)

of tissue absorption, which is independent of metabolism, is identified with the penetration of the apoplasm. Apoplastic fluxes have been even more clearly characterized in measurements of isotopic efflux from tissues. When roots, which for some hours have been permitted to absorb a radiotracer (such as $^{36}Cl^-$), are transferred to a chemically identical (but non-labeled) solution, the pattern of $^{36}Cl^-$ efflux from the tissue can be resolved into three logarithmic phases. These have half-lives ($t_{1/2}$ values) of about 1-2 min, 10 min and 30 h, respectively, for barley roots. The three compartments considered to be responsible for the observed kinetics of isotopic exchange are the apoplasm, the cytoplasm and the vacuole, respectively.

The extracellular tissue space, into which ions appear to enter so readily, is defined by such kinetic experiments as the apparent free space (A.F.S.) of the tissue. When experimenters measure the absorption of tracers by various tissues it is essential to correct for the A.F.S. content because tracer in this compartment has not yet traversed the plasma membrane.

5.2a. WATER FREE SPACE (W.F.S.) AND DONNAN FREE SPACE (D.F.S.).

If, after a short uptake period of the sort shown in Figure 5.1, the radioactively-labeled roots are transferred to well-stirred distilled water there is evident a brief but significant efflux of ^{42}K from the roots. When this is complete, transfer of the roots to a well-stirred non-labeled solution of K_2SO_4 elicits yet another release of tracer as shown in Figure 5.4. The rapidity of these events leads to the conclusion that the radioactive ions are being released from the A.F.S. Futhermore, the above observations seem to indicate that the A.F.S. contains two categories of ions. Those released following immersion of roots in K_2SO_4 solution were bound (adsorbed) to the cell wall and exchanged with the non-radioactive K^+ (see Section 3.4). This fraction is contained within the "Donnan free space" (D.F.S.). The cell wall contains numerous charged groups, particularly the carboxylic groups of pectic substances, which are responsible for binding cations in the D.F.S. Only relatively small quantities of anions are held within this phase. The $^{42}K^+$ released by the initial rinse, represents ions which were completely free to diffuse out of the extracellular volume of the roots. This volume is called the water free space (W.F.S.). The A.F.S. is made up of the W.F.S. and the D.F.S. Note that the A.F.S. is not the actual volume of the apoplasm but rather its equivalent volume based upon its capacity to contain ions of the external solution.

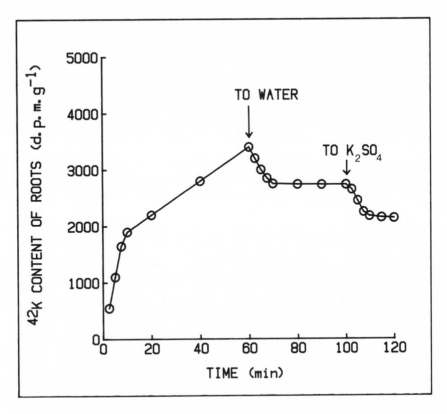

Figure 5.4. Release of $^{42}K^+$ from the apparent free space after transfer to distilled water and subsequently to K_2SO_4 solution. (Glass, unpublished)

In order to estimate the uptake of ions into the symplasm most experimenters rinse the roots after the absorption period in a non-labeled solution which is chemically identical to the labeling solution. By maintaining this "desorption" solution at ~2°C radioactive ions in the D.F.S. and W.F.S. are simultaneously removed, while further absorption into the symplasm is prevented.

In describing bulk flow through soil we defined hydraulic conductivity (Lp) as $J_w/\Delta\Psi$ (Section 3.5). Lp for cell walls of corn roots has been calculated to be 1.4×10^{-9} m s^{-1} bar^{-1} compared to 4×10^{-11} m s^{-1} bar^{-1} for plasma membranes of the same tissue. Clearly the cell wall is about 100 times more conductive to water than the plasma membrane. Permeability coefficients for various inorganic ions in the A.F.S. are also substantially higher (~ 10^{-5} to 10^{-4} m s^{-1}) than corresponding values in the cell membrane (10^{-10} to 10^{-9} m s^{-1}).

To appreciate why cell walls are normally so permeable to ions and water it is necessary to consider wall ultrastructure. Primary cell walls are composed of cellulose microfibrils (10-25 nm in cross-section) embedded in a matrix of more amorphous carbohydrates, (pectins and hemicelluloses). Between the cellulose microfibrils lie spaces (pores) up to 10 nm in diameter, through which hydrated ions (e.g., K^+-diameter 0.54 nm) can pass with relative ease. However, impregnation of the matrix with suberin or lignin reduces wall conductivities (to water and ions) to values corresponding to those of the plasma membrane. Thus, suberization of the epidermis distal from the root tip in corn effectively reduces the entry of phosphate.

DETERMINATION OF THE A.F.S.

The extent of the A.F.S. can be determined most straight-forwardly by extrapolating the slower trans-membrane uptake pattern (shown in Figure 5.3) back to the Y co-ordinate. During the first minutes of the uptake experiment isotopic loss from solution is the combined effect of A.F.S. equilibration and and trans-membrane uptake. The Y intercept at zero time is a good estimate of the A.F.S. isotopic content at equilibrium.

Example: In the experiment shown in Figure 5.3 the initial isotopic content of the uptake solution was equivalent to 500 disintegrations per min (d.p.m) per cm^3 and the A.F.S. was found to contain 50 d.p.m g^{-1} of roots. Clearly, the 50 d.p.m contained within the apoplasm was equivalent to $\frac{50}{500} = 0.1$ cm^3 of the labeling solution. Thus we may say that the A.F.S. was 0.1 cm^3 g^{-1} or approximately 10% of tissue volume. Literature values for A.F.S. rarely exceed 0.20 cm^3 g^{-1} in higher plant tissues.

5.3   Fluxes Across the Plasma Membrane into the Symplasm

Membranes are now universally considered to consist of lipid bilayers in which are dispersed two classes of protein. Intrinsic proteins completely traverse the membrane while extrinsic proteins penetrate only superficially into the bilayer (see Fig. 5.5). The central hydrophobic region of the bilayer (the hydrocarbon tails of membrane phospholipids) represent a considerable barrier to penetration by charged species (Section 2.5). Transport across the bilayer is almost certainly under the control of the intrinsic proteins.

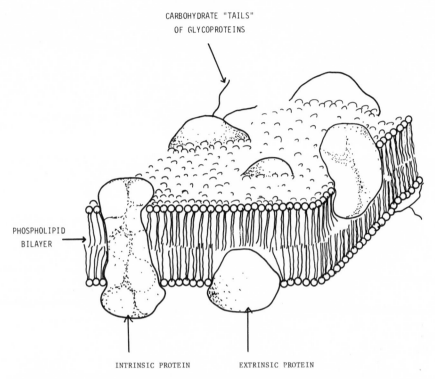

CARBOHYDRATE "TAILS"
OF GLYCOPROTEINS

PHOSPHOLIPID
BILAYER

INTRINSIC PROTEIN   EXTRINSIC PROTEIN

Figure 5.5.   A model of the plasma membrane. The basic structure involves a lipid bilayer in which are embedded intrinsic and extrinsic proteins.

ACTIVE TRANSPORT AND THE NERNST EQUATION

In Section 3.5 it was established that spontaneous (passive) fluxes of matter can only occur down existing gradients of free energy. Charged species (ions) may move passively as a result of driving forces arising from chemical potential differences ($\Delta\mu$) due to concentration asymmetry and/or electrical potential differences (E). Plant membranes typically maintain electrical potential differences (inside negative) of between 100 to 200 mV across the plasma membrane. Such electrical gradients favor the entry of cations but oppose anion penetration.

Consider a root hair extending into soil solution containing 1 mM K^+, with internal K^+ at 100 mM and E amounting to -130 mV. Clearly, the entry of K^+ across the plasma membrane into this cell is energetically "uphill" with respect to the chemical potential difference but "downhill" with regard to the electrical gradient (see Fig. 5.6). A crucial question regarding such ion movements is whether they result from passive or active (biochemical) driving forces. This question is definitively answered by determining whether the ion's distribution can be accounted for exclusively by passive forces. If so, we accept

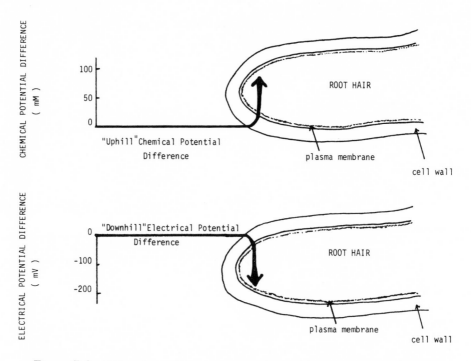

Figure 5.6. Directions of free energy gradients (chemical potential and electrical potential differences) for spontaneous fluxes of K^+ between soil solution and root hair interior. Internal $[K^+] = 100$ mM, external $[K^+] = 1$ mM, membrane electrical potential $= -130$ mV.

the simplest hypothesis, namely, that the flux *is* passive. When the determination yields a negative result the flux must be active because non-equilibrium states can only be sustained by expenditure of energy. In order to address this problem we need to know the direction (resultant) of the chemical and electrical driving forces. The most common method of assessing the magnitude of the passive driving forces arising from chemical and electrical potential differences (the electrochemical potential difference, $\Delta\bar{\mu}$) which act upon an ion species is to make use of the Nernst equation.

5.3a THE NERNST EQUATION

In Section 3.5 we learned that a difference of free energy (chemical potential) could represent the driving force for a diffusional flux of solute. The chemical potential (μ) was given by:

$$\mu = \mu^* + RT \ln C \tag{5.1}$$

where μ^* is the standard state chemical potential, C is the concentration of solute and the terms R and T have their usual meanings. In the case of the root hair we will use subscripts o and i to stand for inside and outside terms. The chemical potential inside is given by:

$$\mu_i = \mu^* + RT \ln C_i \qquad (5.2)$$

Similarly, $\mu_o = \mu^* + RT \ln C_o$. $\qquad (5.3)$

The difference of μ ($\Delta\mu_{oi}$) is given by:

$$\Delta\mu_{oi} = RT \ln C_o - RT \ln C_i \qquad (5.4)$$

However we noted that the root hair membrane was electrically polarized ($E = -130\text{mV}$) and this strongly influences the free energy of a charged solute. Indeed, Equation 5.1 must include a term for the electrical potential (Ψ) which exists in each compartment although, as will become evident, we will be more interested in the difference of electrical potential ($\Delta\Psi$ or E) between the two compartments. Thus, Equation 5.1 becomes:

$$\bar{\mu} = \bar{\mu}^* + RT \ln C + z\, F\Psi \qquad (5.5)$$

We distinguish μ (the chemical potential) from $\bar{\mu}$ (the combined electrical and chemical or electrochemical potential); $\bar{\mu}^*$ is the standard state electrochemical potential, z is the charge on the ion ($+1$ for K^+, -1 for NO_3^-), F is the Faraday constant ($96.5 \text{ kJ mol}^{-1} \text{V}^{-1}$) and Ψ is the electrical potential (usually measured in mV).

Hence when an electrical potential difference exists (in the root hair this was a difference of 130 mV) the resultant electrochemical potential difference between the two compartments (cf 5.4) is given by

$$\Delta\bar{\mu}_{oi} = (RT \ln C_o + z\, F\Psi_o) - (RT \ln C_i + z\, F\Psi_i) \qquad (5.6)$$

Now, if $\bar{\mu}_i > \bar{\mu}_o$ then K^+ might be expected to move passively out of the cell. If $\bar{\mu}_o > \bar{\mu}_i$ then K^+ should enter the cell. If $\bar{\mu}_o = \bar{\mu}_i$ then there should be no flux in either direction since there is no passive driving force on the ion. Under these conditions:

$$(RT \ln C_o + z\, F\Psi_o) = (RT \ln C_i + z\, F\Psi_i) \qquad (5.7)$$

Gathering similar terms on each side of the equation, Equation 5.7 becomes:

$$z\, F\Psi_i - z\, F\Psi_o = RT \ln C_o - RT \ln C_i \qquad (5.8)$$

and

$$z\, F(\Psi_o - \Psi_i) = RT \ln \frac{C_o}{C_i} \qquad (5.9)$$

Hence

$$(\Psi_o - \Psi_i) = \frac{RT}{z\, F} \ln \frac{C_o}{C_1} \qquad (5.10)$$

if we replace $(\Psi_i - \Psi_o)$ by E^N (the Nernst electrical potential difference), then at equilibrium

$$E^N = \frac{RT}{z\,F} \ln \frac{C_o}{C_1} \tag{5.11}$$

Substituting values for constants at a temperature of 20°C in Equation (5.11) we obtain

$$E^N = \frac{58}{z} \log_{10} \frac{C_o}{C_i} \tag{5.12}$$

It is important to remember that this equation is based upon the assumption (Equation 5.7) that $\bar{\mu}_i = \bar{\mu}_o$ (i.e. that $\Delta\bar{\mu}_{oi} = 0$). Hence the value of E^N predicts the electrical potential different (in mV) when the resultant chemical potential difference is just balanced by the resultant electrical potential difference.

In the case of the root hair (Section 5.3a) we can calculate E^N to be

$$\frac{58}{+1} \log_{10} \frac{1}{100} \text{ or } -116 \text{ mV.}$$

In order to satisfy the criterion that $\Delta\bar{\mu}_{io} = 0$ (equation 5.7) it would therefore require a potential difference of -116 mV to just balance the existing concentration asymmetry, (99 mM) across the membrane. Since E was observed to be -130 mV, the required electrical potential difference is just adequate. We, therefore, accept that the distribution of K^+ is satisfactorily accounted for in terms of the existing free energy gradients. Although the entry of K^+ appears to be "uphill" according to the chemical potential difference, this is counteracted by the "downhill" electrical potential difference. The resultant driving force is *into* the cell. Thus we conclude that K^+ entry into this cell is by passive transport. If, however, the measured E had been, for example, -75 mV this value would have fallen far short of the required E^N and we would then have concluded that the electrical potential difference was insufficient to sustain the gradient of K^+. In so doing we would invoke active mechanisms to account for the distribution of K^+. In fact any deviation between E (measured) and E^N (the equilibrium potential difference) indicates that the system is not at equilibrium because the derivation of E^N is predicated upon the assumption that $\Delta\bar{\mu} = 0$ (Equation 5.7). Hence the value of $E-E^N$ is a measure of the deviation from equilibrium and can be used to calculate $\Delta\bar{\mu}$ from the relationship:

$$\Delta\bar{\mu} = zF(E - E^N) \tag{5.13}$$

In the above case, where E was speculated to be -75 mV,

$$\Delta\bar{\mu} = +1 \times 96.5 \,(-75 -(-116))$$

$$\Delta\bar{\mu} = 3956.5 \text{ Joules mol}^{-1}$$

The positive value indicates that K^+ is at a higher electrochemical potential (free energy) inside the cell. The entry of K^+ into such a cell must therefore be active. Having elaborated upon a definitive criterion for active transport it should now be possible to state precisely which inorganic nutrients are actively and which are passively absorbed by plant tissues. Yet, surprisingly, the literature contains numerous contradictions in this regard, particularly with regard to K^+.

The Nernst criterion for active transport was applied to the distribution of several cations and anions in oat and pea tissue by Higinbotham and his coworkers in 1967. For each ion, the external concentration was maintained at 1 mM, membrane potentials were measured and E^N calculated from the equation. The results (Table 5.2) revealed that none of the anions were distributed according to their electrochemical potential gradients. For pea roots, Ca^{2+}, Mg^{2+} and Na^+ were present at lower tissue concentrations than predicted from E^N, suggesting extremely low membrane permeability for these ions

Table 5.2. Use of the Nernst equation to predict internal concentrations of various ions in pea and oat roots. The authors substituted the measured value of E into Equation 5.11 to calculate C_i values. The predicted values of C_i obtained are therefore based upon equilibrium distributions of each ion. External ion concentrations (C_o) were 1mM. E for pea roots was -110 mV, for oat, -84mV. (From Higinbottom *et al.*, *Plant Physiology*, 42:37-46, 1967)

| Ion | Ion concentrations in tissues (mM) | | | |
| | Pea roots | | Oat roots | |
	Predicted conc.	Measured conc.	Predicted conc.	Measured conc.
K^+	74	75	27	66
Na^+	74	8	27	3
Mg^{++}	2,700	3	350	17
Ca^{++}	10,800	2	1,400	3
NO_3^+	0.0272	28	0.0756	56
Cl^+	0.0136	7	0.0378	3
$H_2PO_4^+$	0.0136	21	0.0378	17
$SO_4^=$	0.000094	19	0.00071	4

or active extrusion mechanisms. Only K^+ was at equilibrium with respect to $\Delta\bar{\mu}_{K^+}$ For oats the situation was practically identical except that K^+ was actively absorbed.

Two very clear conclusions emerge. 1. Since membrane potential differences are negative inside, the absorption of an anion is always active if tissue concentration exceedes external concentration. This is true because under these conditions absorption is against both electrical and chemical potential differences. 2. The situation for cations can only be evaluated after careful evaluation of the concentration and electrical terms which apply. Higinbotham's data were derived using 1 mM external ion concentrations. Authors who extrapolate from these particular conditions to a universal picture are in error. For each case it is critical to know external ion concentrations. Consider a situation where $[K^+]_i = 100$ mM, $[K^+]_o = 1$ mM and measured $E = -120$ mV.

E^N can be calculated to be -116 mV and we accept a passive distribution of K^+. However if $[K^+]_o$ were reduced to 0.1 mM then E^N becomes -174 mV and we must invoke active transport to account for K^+ uptake.

One more ion, the H^+ ion (or proton), will be considered because of its importance to cell metabolism, in the maintenance of cell pH, its contribution to membrane potentials and its possible involvement in the chemiosmotic transport of other ions (Section 5.7).

Plant roots are capable of substantial H^+ extrusion. Under certain circumstances this H^+ efflux is related to the balance between cation and anion absorption (Chapter 7). For example, when cation uptake exceeds anion uptake as for example in K_2SO_4 solution, the external solution becomes acid. Figure 5.7 shows the rapidity of pH changes due to K^+ uptake from K_2SO_4 by roots of barley. With cytoplasmic pH at about 7 (10^{-7} mol $H^+ l^{-1}$), external pH at 5.3 ($10^{-5.3}$ mol $H^+ l^{-1}$) and a membrane potential of about -100 mV we can calculate $\Delta\bar{\mu}_{H^+}$ from equation 5.13. If H^+ were at equilibrium then

$$E^N \text{ would} = 58 \log \frac{10^{-5.3}}{10^{-7}}$$

$$= +98.6 \text{ mV}$$

To maintain this asymmetric $[H^+]$ passively demands that E be $+98.6$ mV. Since the measured E was -100 mV, H^+ distribution is far from equilibrium and

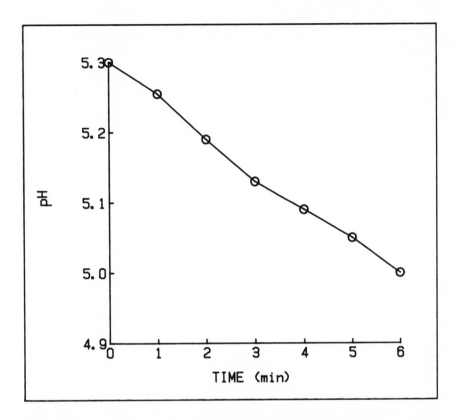

Figure 5.7.   pH changes of the external solution associated with K⁺ uptake by roots of a barley variety (Fergus). Roots were immersed in a solution containing 500 μM K_2SO_4, and 500 μM $CaSO_4$, at an initial pH of 5.3. (From Glass *et al.*, *Plant Physiology*, 68:457-459, 1981)

$$\Delta\mu_{H^+} = 96.5\,(-100 - 98.6)$$
$$= -19{,}060 \text{ J mol}^{-1}$$

This is an extremely large $\Delta\bar{\mu}$, whose −ve value indicates that H⁺ is at a lower electrochemical potential inside the cell. The observed H⁺ efflux must therefore involve active driving forces. Moreover, in considering this large passive driving force for H⁺, which is in a direction from outside the root toward the inside, it is quite remarkable that the cytoplasm is able to maintain its pH value at close to 7 even when external pH is quite low. In hydroponic systems, for example, barley plants are able to grow quite well at pH values below 5. Practical aspects of acid tolerance in crop plants are considered in Chapter 9.

5.4   Origins of Membrane Electrical Potentials

The origins of bio-electrical phenomena have intrigued scientists since the late eighteenth century when Luigi Galvani discovered that an electrical discharge could cause muscle contraction in frogs' legs. Until quite recently it was believed that trans-membrane potential differences were generated passively as ions diffused along their concentration gradients. Of course it was acknowledged that active transport might be responsible for generating the concentration gradient in the first place but this dependence upon metabolism was indirect. Today it is accepted that there may also be contributions to E arising directly from active fluxes of ions, the so-called electrogenic potentials.

DIFFUSION POTENTIALS (E_d)

To generate diffusion potentials requires that a gradient of ion concentration exists across a membrane that is differentially permeable to the cation and anion species present. Consider a membrane which is more permeable to cations than to anions. Imagine that this membrane separates two solutions of KNO_3, respectively 100 and 1 mM in concentrations, as shown in Figure 5.8.

The gradient of concentration will cause K^+ and NO_3^- to diffuse from A to B but, because the membrane is more permeable to K^+, a

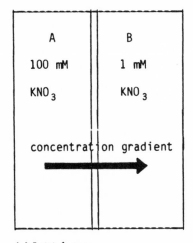
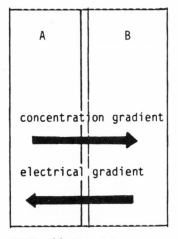

(a) Initial state   (b) Equilibrium state

Figure 5.8.   A hypothetical diffusion potential

small charge separation will develop across the membrane (A becoming $-$ve by reference to B) as K^+ migrates more readily than NO_3^-. The resulting potential difference will retard K^+ diffusion from A to B and at equilibrium the potential difference (the diffusion potential, E_d) will just counteract the tendency for K^+ to move from A to B. The magnitude of E_d will be a function of the ion concentration difference and the relative permeabilities to cations and anions present. The actual value of E_d can be predicted from the Goldman equation:

$$E_d = \frac{RT}{F} \ln \frac{P_{K^+}[K^+]_B + P_{NO_3^-}[NO_3^-]_A}{P_{K^+}[K^+]_A + P_{NO_3^-}[NO_3^-]_B}$$

Contributions to $E_{(d)}$ arising from other ions present may be evaluated by adding terms for their concentrations and membrane permeabilities to the equation. If the membrane were completely impermeable (i.e. $P = 0$) to NO_3^- then the equation would simplify to the Nernst equation for K^+. This is the case for ion-specific electrodes such as the pH electrode. The bulb of the pH electrode is made of special H^+-permeable glass. The pH meter is simply a voltmeter which measures the diffusion potential generated across the bulb.

If the maintenance of a concentration gradient depends upon metabolism we might anticipate that any interruption of energy supply would gradually cause the gradient to come to equilibrium. Many workers have provided evidence that under certain conditions perturbation of metabolism (by temperature, light or metabolic inhibitors) creates an immediate change of E which is too rapid to be due to effects upon concentration differences or P values. This and other evidence suggests the existence of another source of E, referred to as electrogenic potentials (E_P)

ELECTROGENIC POTENTIALS (E_p)

Figure 5.9 demonstrates the effect of KCN upon the transmembrane potential of sunflower roots. Note that within minutes of its application this metabolic inhibitor reduced E by about 90 mV. It is particularly interesting that the inhibited value of E was quite close to the calculated diffusion potential based upon the distribution of K^+. Research workers often accept results based upon metabolic inhibitors with some misgivings because of uncertainties concerning side-effects. The inhibitor might, for example, be changing membrane permeability in addition to halting respiration. Figure 5.10 shows the effect of light on E in a single cell of *Nitella*. Again the diffusion potential for K^+ (E_K) is provided for reference. In this case, providing

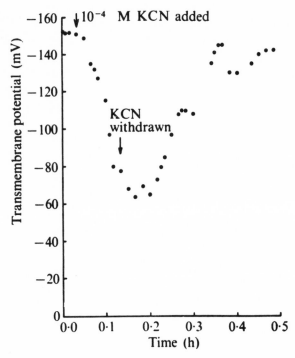

Figure 5.9. Effect of KCN upon trans-membrane electrical potential differences (*E*) in roots of sunflower. (From Graham and Bowling, *Journal of Experimental Botany*, 28:886-893, 1977)

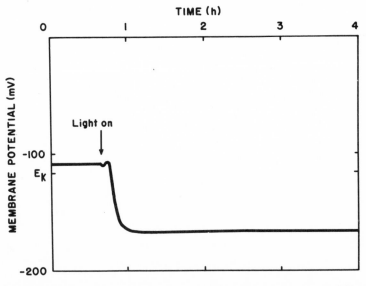

Figure 5.10. Effect of light upon the trans-membrane potential differences in Nitella. (From Spanswick, *Canadian Journal of Botany*, 52:1029-1034, 1974)

a light source to a photosynthetic cell hyperpolarized the cell membrane by roughly 60 mV. There are two points to be stressed about these data. Firstly, the measured E values are more negative than would be predicted from E_d. Unless we are overlooking some ions which make a significant contribution to E_d it would appear that there is an additional source of E. Secondly the rapidity of the reduction of E to values close to E_d suggests that this additional source of E is directly dependent upon metabolism. It is now believed that the active transport of particular ions across the plasma membrane may contribute significantly to charge separation and the generation of E. Clearly, if cations and anions are transported in the same direction in equal amounts then there is no charge separation. However if there are unbalanced cation or anion fluxes then charge separation is generated. The extrusion of H^+ from the cell is believed to be an essentially universal phenomenon associated with diverse processes (to be discussed subsequently) and this flux may generate a large electrogenic potential (E_p). We may reduce the various membrane potentials and their sources and pathways to an equivalent circuit of the sort used by electricians. Figure 5.11 shows such a circuit.

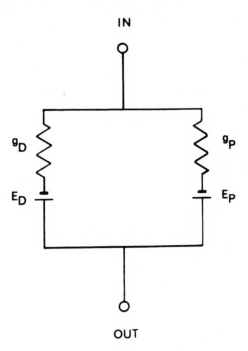

Figure 5.11. An equivalent circuit for showing the pump emf (E_p) and conductance (g_p) and the diffusion potential (E_D) and conductance (g_D). (From Spanswick, *Annual Review of Plant Physiology*, 32:267-289, 1981)

The circuit is seen to consist of two paths in parallel, with E values corresponding to diffusion (E_D) and electrogenic potential (E_p) differences and conductances through the diffusion (g_D) and pump (g_p) channels, respectively. Besides explaining the origin of E the preceding discussion provides the background to understand an extremely important concept in ion transport, namely, chemiosmotic transport mechanisms which are discussed in Section 5.7.

5.5   Mechanisms for Transport Across Membranes: The Carrier Concept

The designation active or passive in reference to ion transport across membranes simply identifies the immediate source of energy responsible for the flux. If passive, the source of energy is to be found in the electrochemical forces which act upon the ion, even though part of the latter (e.g. E_p) may be metabolically derived. If active, there is obviously a need for more direct coupling between energy supply and the transport process. Despite these designations few scientists believe that even passive fluxes of ions occur by diffusion; the cause of their doubt is the relatively impermeable nature of the membrane to all but a few nutrients (e.g. NH_3). The high resistance to diffusion presented by the lipid bilayer clearly represents a severe constraint upon nutrient uptake and consequently to growth. Thus it is currently held that both active and passive fluxes probably involve interaction with membrane proteins which facilitate transport. In this section we explore the properties of the "membrane components" responsible for trans-membrane fluxes.

Largely through the considerable efforts of Epstein, working at Davis, California, it is now widely held that inorganic ions traverse cell membranes by first binding with membrane proteins (carriers), which are thought to have the same specificity for individual ions as enzymes have for their substrates. The evidence for this carrier hypothesis is based upon the similarity between curves for ion absorption versus external ion concentration (Fig. 5.12) and those for enzyme activity versus substrate concentration. The dependence of enzyme activity upon substrate concentration, first modelled by Michaelis and Menten (1913), may be described by the equation:

$$v = \frac{V_{max} \ [S]}{K_m + [S]}$$

where v = velocity of enzyme action, V_{max} is the maximum velocity,

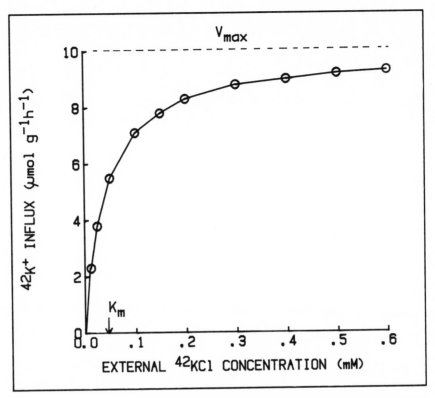

Figure 5.12. A plot of $^{42}K^+$ influx, as a function of external ^{42}KCl concentration, into roots of intact corn plants. Experiment conducted at 30°C in solutions containing 500 μM $CaSO_4$. (Glass, unpublished)

[S] is the substrate concentration and K_m is the substrate concentration giving half maximum velocity. This equation (the Michaelis Menten equation) is derived from a model of enzyme activity in which the enzyme combines with substrate to form a complex that breaks down to yield enzyme plus products:

$$E + S \rightleftharpoons ES \rightleftharpoons E + P$$

With a fixed quantity of enzyme and low concentrations of substrate ([S]), conditions in which enzyme velocity is limited by substrate availability, there is a strictly linear response to increasing [S]. However velocity does not increase indefinitely with increasing [S]. As enzyme activity becomes saturated a maximum velocity (V_{max}) is approached asymptotically. At this [S] all enzymes are working at full capacity. Note, this model also describes the pattern of ion binding to soil particles (Section 3.4).

By analogy with this enzyme model, Epstein proposed the carrier concept of membrane transport, in which the essential features of substrate recognition and membrane transport were attributed to membrane carriers. According to this model the first stage in membrane transport involves binding of the ion to a specific carrier. This ion-carrier complex is thought to traverse the membrane, releasing the ion to the cytoplasm. Epstein further suggested that the Michaelis Menten 'constants', K_m and V_{max}, might be used to describe the kinetics of transport. A prediction of the model was that specific carriers would transport particular ions and that these carriers might fail to discriminate between ions of similar charge and dimension. These predictions were borne out; plant tissues are unable to distinguish between ion pairs such as Br^- and Cl^-, K^+ and Rb^+, SO_4^{2-} and SeO_4^{2-}. By analogy with enzyme kinetics, Epstein referred to these pairs of ions as competitive inhibitors. In mixtures they are thought to compete for the carriers' binding sites, thus mutually reducing the uptake of the other ion.

Initially the investigations of Epstein and his co-workers concentrated upon absorption from relatively dilute solutions (< 1 mM). Subsequently, Epstein and, later, other groups began to investigate patterns of absorption at much higher concentrations, even beyond 50 mM. In the concentration range above 1 mM ion fluxes revealed complex concentration dependence inconsistent with the operation of a single Michaelis Menten system (Fig. 5.13). Moreover, these high concentration fluxes were qualitatively different from those < 1 mM. In the 1960's and 70's there was considerable controversy regarding the interpretation of these high concentration kinetics.

Epstein (see *Review*, 1976) argued that they were due to a second class of carriers located in the plasma membrane; Laties preferred the tonoplast as the site of their activity; Per Nissen proposed that both low and high concentration fluxes were due to a single transport system which underwent concentration-dependent phase changes, while other workers believed that the high concentration fluxes were the result of diffusion across the plasma membrane.

Recently the question has been the subject of renewed attention. Kochian and Lucas (1982), using corn roots showed that K^+ uptake increased in a strictly linear fashion in the high concentration range (Figure 5.14). These fluxes have many of the characteristics of those which occur through the transmembrane K^+ channels of animal membranes.

Ionic channels were first detected, on the basis of their electrical properties, in the membranes of nerve cells, where they facilitate the massive Na^+ and K^+ fluxes responsible for nerve impulses. They will

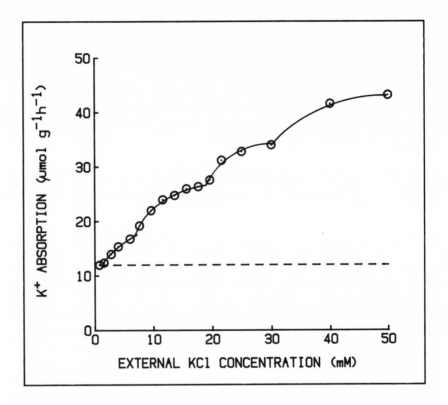

Figure 5.13. Rates of K^+ absorption by excised barley roots as a function of external KCl concentration. Dotted line shows the maximal rate of absorption, V_{max}, of the low concentration uptake mechanism. (From Epstein and Rains, *Proceedings of the National Academy of Sciences*, 53:1320-1324, 1965)

probably be found to be present in all cell membranes. Although a detailed treatment of channels requires much more biophysical background than is assumed in this book, the student should be aware of their existence and their properties, since channels will probably be discovered to participate in many important regulatory functions in plants. Thus far they have been detected in several plant cells including guard cells of broad bean (*Vicia faba*), several different algae, and in protoplasts from wheat and from corn roots.

Basically channels are macromolecular pores which traverse cell membranes. Ions can move very rapidly through a single channel ($> 10^6$ ions per sec) and always down the electrochemical potential gradient. Hence they are not to be confused with transporters which are capable of moving solutes against their gradient. Although channels are considered to be pore-like they should not be viewed as open, non-specific pores. By virtue of their dimension and shape, together with specific mechanisms causing opening and closing, they can be quite specific in the solutes they permit to enter or leave the cell. Hence they have often been named according to the ion they allow to pass through the cell membrane, e.g., K^+ channels, or Ca^{2+} channels. Among factors responsible for opening/closing channels, membrane electrical potentials and various hormones can act as regulatory signals.

Although the controversy regarding high concentration fluxes has not been settled to everyone's satisfaction, greater emphasis, in recent years, has been directed towards understanding biochemical and biophysical aspects of membrane transport. In this regard, the chemiosmotic theory of energy transduction (Section 5.7) has exerted a powerful influence on contemporary thinking.

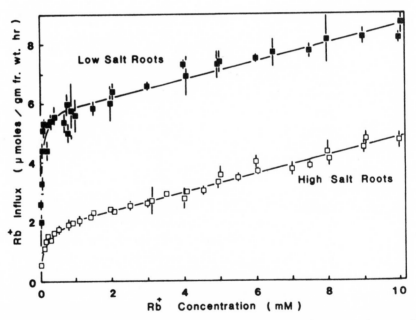

Figure 5.14. $^{86}Rb^+$ influx into corn root segments as a function of external concentration and salt status. Plants were grown in 0.2 mM $CaSO_4$ (low salt roots) or 0.2 mM $CaSO_4$ plus 5 mM KC1 (high salt roots). The authors interpreted the observed kinetics as due to additive effects of two separate transport systems; one saturable system reaches its maximum activity at $\sim$ 1 mM Rb^+, while a second system continues to increase linearly up to 10 mM Rb^+ (Kochian and Lucas, *Plant Physiol.* 70:1723-1731, 1982).

5.6   ATPases: The Link with Metabolism

We have established on the basis of thermodynamic considerations that certain ion fluxes may be active. This conclusion is corroborated by the often reported reductions of ion fluxes by treatments which block metabolism. Inhibitors which prevent electron transport, for example, cyanide (CN^-), or prevent ATP synthesis, e.g. arsenate or dinitrophenol (DNP), have been widely used. Again, algal cells have been particularly useful experimental tools because of the convenience with which photosynthetic energy supply can be manipulated. What emerges from these studies is the conclusion that, in general, ion transport may be energized by respiration or by photosynthesis depending upon the prevailing conditions. In root tissue, naturally, respiration is the source of energy. For example in barley roots the absorption of K^+ is extremely sensitive to the availability of oxygen in the uptake solution (see Fig. 5.15).

Although we appreciate that the energy for active transport comes ultimately from respiration (in roots) or photosynthesis (in leaves and single-celled algae), the details of the linkages between these processes and ion transport are only just beginning to be clarified (see also Section 7.4). Are the transport process powered by energy derived

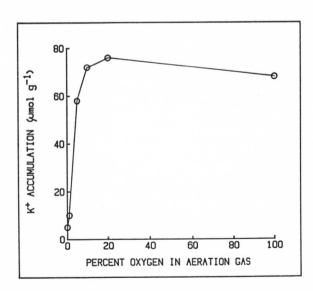

Figure 5.15.   Effect of oxygen in the gas stream supplied to excised barley roots on the accumulation of K^+. (From Hoagland and Broyer, *Plant Physiology*, 11:471-507, 1936)

directly from the reduction/oxidation (redox) reactions of the electron transport chain (e.t.c.) or indirectly via ATP? One of the earlier models of active transport, proposed by Lundegårdh (1940), involved a direct linkage between electron flow and ion transport. The model was based upon the observed stimulation of respiration (termed "anion respiration") which occurred when roots were placed in salt solution. Lundegårdh suggested that oxidized cytochromes were capable of binding anions and passing them along the cytochrome chain in a direction opposite to electron flow. Unfortunately this hypothesis demanded an orientation of the e.t.c within the plasma membrane. The subsequent realization that cytochromes were localized within membranes of chloroplasts or mitochondria made this model untenable.

Other schemes of a basically similar sort have been proposed since Lundegårdh's and the results of some inhibitor studies have been interpreted as supporting a direct coupling between ion uptake and electron flow. However in the main it is generally considered that this is not the means whereby energy is provided for ion transport across the plasma membrane. Nevertheless, as will become evident (Sec. 5.7), fluxes of H^+ across the inner membranes of chloroplasts and mitochondria are indeed powered by direct coupling to electron transport.

Evidence supporting the role of ATP as the source of energy for ion transport across the plasma membrane is derived from two sources:-

(1) THE EXISTENCE OF MEMBRANE ATPases.

Membrane-associated enzymes capable of hydrolyzing ATP (ATPases) have been demonstrated cytologically and have been isolated from numerous plant tissues since the late 60's and early 70's. A major difficulty however, has been to demonstrate that these ATPases are specifically related to membrane transport. Moreover, when ATPase activity is assayed *in vitro*, following cell disruption and fractionation, there exists some uncertainty concerning the exact source of the resulting membrane fragments (do they originate from plasma membrane, tonoplast, or endoplasmic reticulum?). At one time, the presence of glucan synthetase activity and characteristic staining properties were advanced as evidence of plasma membrane origin of membrane fragments but unfortunately the validity of these criteria has now been seriously questioned. The major criterion for an involvement in transport has been to establish that the ATPase activity is stimulated (*in vitro*) by addition of specific ions. For example, Na^+/K^+ ATPases of animal plasma membranes, which mediate the linked counter-fluxes of Na^+ and K^+ (Na^+ out/K^+ in), are synergistically stimulated by the presence of Na^+ and K^+. There are, indeed, many reports of K^+-stimulated membrane ATPases, isolated from cereal roots. In addition, indirect evidence for the involvement of these enzymes in K^+ absorption comes from the observed correlations between rates of K^+ absorption and in vitro ATPase activity (Fig. 5.16).

(2) METABOLIC INHIBITORS

Inhibitors such as arsenate which compete with Pi and thus reduce ATP formation have been reported to inhibit the uptake of Cl^-, Pi and SO_4^{2-} in corn roots, while actually stimulating electron transport and O_2 use (Weigl, 1964). This demonstration strongly argues for an involvement of ATP, rather than electron transport *per se*, in ion absorption. In a limited number of cases where ATP levels have been manipulated (e.g. by use of inhibitors) good correlations have been observed between ATP levels and ion fluxes. For example,

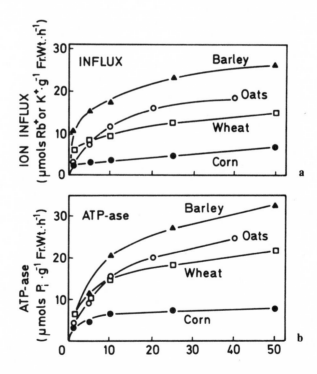

Figure 5.16.   (a) The effect of KCl or RbCl concentrations on the influx of K^+ into oat roots and the influx of Rb^+ into roots of barley, corn and wheat. (b) The effect of KCl or RbCl concentrations of the (K^+ or Rb^+-stimulated) ATPase activities in membranes isolated from roots of these species. (From Fisher et al., *Plant Physiology*, 46: 812-814, 1970)

when ATP levels were reduced by CN^- treatment in *Neurospora* (the bread mold) the electrogenic potential (E_p), due to active H^+ efflux, was correspondingly reduced (Slayman et al., 1973). Similarly, in red beet influxes of Cl^- and K^+ were demonstrated to be correlated with tissue ATP levels when the latter were reduced by various respiratory inhibitors (see Fig. 5.17). Thus, both influx and efflux of certain ions appear to depend upon available ATP levels.

These studies bring us closer to an understanding of the role of ATP in ion absorption but they fail to reveal whether ATP is a direct source of energy for transport or whether it brings about energy-dependent fluxes by indirect means. To appreciate current perspectives on this most important topic we must introduce chemiosmotic principles into our arguments.

5.7 CHEMIOSMOTIC THEORY AND COUPLED SOLUTE FLUXES

The free energy changes associated with the redox reactions of the e.t.c. of mitochondria and chloroplasts are conserved in the synthesis of ATP. The crucial question which was addressed by Peter Mitchell (author of the Chemiosmotic Theory) was the nature of the link between electron transport and ATP synthesis. After years of fruitless searching for a biochemical intermediate to link these two processes, by numerous research groups, Mitchell proposed that the link was a proton gradient. Specifically, he suggested that electron transport within energy transducing membranes was directly coupled to vectorial H^+ pumping across these membranes.

As a result, during electron transport an electrochemical potential difference ($\Delta \bar{\mu}_{H^+}$) is generated across the transducing membrane. This $\Delta \bar{\mu}_{H^+}$, which Mitchell called the proton motive force, was due to the gradient of H^+ (ΔpH) and E (the electrical gradient across the membrane). Mitchell further proposed that the free energy associated with $\Delta \bar{\mu}_{H^+}$ was used to drive the endergonic reaction of ATP synthesis as protons returned (down their free energy gradient) via membrane ATPases. Actually ATPases are reversible enzymes which should be called ATP synthetases when they operate in reverse. It was soon established that H^+ were, indeed, transported into the thylakoid cavity during the light reactions of photosynthesis and that chloroplasts could bring about ATP synthesis in the dark if a pH gradient were artifically set up across the thylakoid membrane. Supporting evidence for Mitchell's ideas was also forthcoming from mitochondrial studies. Mitchell considered that $\Delta \bar{\mu}_{H^+}$ (or in fact $\Delta \bar{\mu}$ for any ion) represents a source of free energy which may be used to achieve

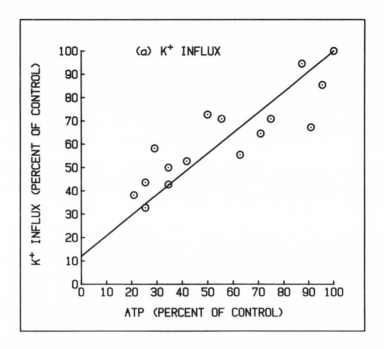

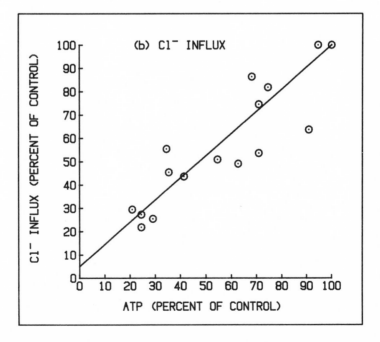

Figure 5.17. (a) K⁺ influx and (b) Cl⁺ influx as a function of cellular ATP levels in red beet. ATP levels were reduced by application of KCN. (From Petraglia and Poole, *Plant Physiology*, 65:969-972, 1980)

biochemical work (e.g. ATP synthesis) or biophysical work (e.g. ion transport) provided an appropriate coupling mechanism exists to link $\Delta\bar{\mu}_{H+}$ and the particular endergonic reaction which is driven by $\Delta\bar{\mu}_{H+}$. In chloroplasts and mitochondria (and, incidentally, in the outer membranes of prokaryotic cells) $\Delta\bar{\mu}_{H+}$ is generated by redox reactions. In other membranes $\Delta\bar{\mu}_{H+}$ may result from H^+ pumping ATPases operating at the expense of ATP.

It must be stressed that active H^+ efflux, at the expense of ATP hydrolysis, represents the primary transport process. Coupling between $\Delta\bar{\mu}_{H+}$ and secondary transport reactions might occur by several different means:

a. **Electrogenic Coupling**: in which membrane hyperpolarization (due to active H^+ efflux) has the effect of increasing the passive driving force for cation or anion movement.

b. **Chemical Coupling**: in which there is an obligate linkage between H^+ pumping and cation movement in the opposite direction (counter-transport or antiporting) or anion movement in the same direction (co-transport or symporting). This type of coupling, unlike that in a., can achieve active transport of the secondary solute. Another form of chemical coupling makes use of $\Delta\bar{\mu}_{H+}$ to symport cations, anions, or even organic solutes such as sugars and amino acids. These examples are shown in Figure 5.18.

You will recall that roots have a well developed capacity for active H^+ efflux and that E is commonly more negative than predicted, presumably because of E_p. Indeed, earlier workers had proposed schemes in which cation and anion uptake occurred in exchange for H^+ and OH^-, respectively. Robertson (1967) proposed that the separation of protons and electrons is a fundamental biological process which might bring about the movement of other ions. However the chemiosmotic theory has provided a powerful conceptual base which brings together and rationalizes diverse transport phenomena. Moreover, because of the controversy it evoked initially, the theory has been subjected to intensive experimental investigation. Not surprisingly, then, chemiosmotic principles have been enthusiastically embraced in recent years and applied to numerous plant systems. In 1970 F. A. Smith proposed that Cl^- transport in *Chara* might be linked to pH gradients at the plasma membrane. Hodges, in 1973, suggested that membrane ATPases might generate $\Delta\bar{\mu}_{H+}$ across root plasma membranes, thus increasing the passive driving force for cation uptake. The same author suggested that H^+ extrusion would simultaneously lead to an OH^- (pOH) gradient across the membrane

(a)   ELECTROGENIC COUPLING

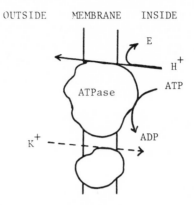

(b)   DIRECT (CHEMICAL) COUPLING

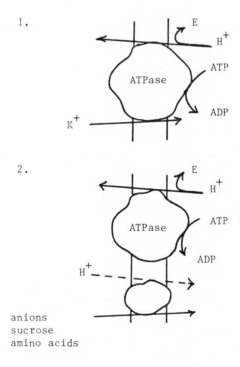

Figure 5.18.   Possible mechanisms of coupled solute fluxes involving H^+ transport. (a) membrane hyperpolarization increases the passive driving force for K^+ entry (b) solute influx is obligatorily linked to the active H^+ flux (antiporting as in b.1) or to the passive return of H^+ along the gradient of $\Delta\bar{\mu}_{H^+}$ (symporting as in b.2).

which might bring about OH⁻/anion counter transport. Thus, according to this model the cation-stimulated ATPase is directly responsible for cation uptake and indirectly responsible for anion uptake.

Strong support for chemiosmotic K^+ transport has come from the work of Marré and his associates. In response to fusicoccin, a toxin produced by the phytopathogenic fungus *Fusicoccum amygdali*, plant tissues demonstrate increased H^+ efflux, membrane hyperpolarization and increased K^+ influx; all events which are consistent with an electrogenic transport of K^+ along its $\overline{\varDelta\mu}_{K^+}$. However, there is by no means total agreement on this point and several workers have challenged this interpretation. Recently, evidence has been presented in several papers to suggest that various anions, (NO_3^-, SO_4^{2-} and $H_2PO_4^-$) and even K^+ may be absorbed via symport with H^+ along the gradient of $\overline{\varDelta\mu}_{H^+}$.

Before 1980 most of the evidence for electrogenic proton translocation via membrane ATPases was indirect. Since this time, however, considerable progress has come about through the use of partially purified membrane fractions to characterize these ATPases biochemically. When whole tissues are ground in buffered solutions the disrupted membrane systems reseal to form spherical vesicles which can be isolated and partially purified by differential centrifugation.

Using vesicles prepared from different plant sources (e.g. tobacco callus, pea stems, and corn roots) various workers have been able to establish the capacities of these membranes to generate a proton motive force ($\overline{\varDelta\mu}_{H^+}$) when provided with ATP. It has been shown that plant cells possess at least two distinct kinds of membrane ATPases. These are identified by their separate locations in the plasma membrane and the tonoplast, respectively, and by their differential sensitivity to inhibitors. For example the plasma membrane ATPase is inhibited by vanadate while the tonoplast ATPase is insensitive to this compound. Surprisingly, nitrate ions inhibit the (in vitro) activity of the tonoplast ATPase, although this ion fails to inhibit the plasma membrane ATPase. In addition to these proton translocating ATPases, a tonoplast H^+ pump which uses pyrophosphate as its source of energy (rather than ATP) has recently been described.

Various methods have been employed to determine the molecular weights of these ATPases which appear to be quite large molecules (M.W. $\simeq$ 100 kD). The tonoplast ATPase has recently been described as composed of 3 subunits; a 16 kD sub-unit which spans the lipid bilayer and is thought to constitute the proton channel, a 72 kD unit responsible for ATP hydrolysis and a 60 kD unit which may be responsible for regulating the activity of the whole complex.

In summary, chemiosmotic mechanisms represent an exciting prospect for explaining the energy-dependence of ion transport. In the coming years the details of the primary transport reactions (transmembrane H^+ transports) will surely be clarified. Notwithstanding the importance of these reactions, the coupling between the resulting energy source $(\varDelta\bar{\mu}_{H+})$ and the transport of the various solutes required by plant cells remains as an exciting prospect for study. As will be explained in Chapter 7, the absorption of each ion appears to be independently regulated according to the plant's requirements for that nutrient. Therefore, although $(\varDelta\bar{\mu}_{H+})$ may represent the driving force for secondary transport reactions, the regulation of the carriers responsible for individual solute fluxes must represent a critical process in determining the extent of these fluxes.

Summary

The unique inorganic composition of plant cells is achieved through the activity of specific ion transporters located in an otherwise impermeable plasma membrane. Transport across particular membranes (influx or efflux) can be defined as active only after an assessment of the combined passive driving forces (chemical and electrical potential differences) which may act upon the ion. If the ion's distribution can not be accounted for by the passive driving forces, active transport is invoked. Typically plant membranes sustain electrical potential differences of about -100 to -200 mV across plasma membranes as a result of passive diffusive fluxes and active electrogenic fluxes.

The isolation of membrane ATPases and the documented correlations between ion fluxes and ATPase activity in specific tissues suggest that ATP is the energy source for ion transport. It is now widely held that the primary transport reaction in which ATP is consumed is the transport of H^+ across cell membranes by H^+ pumping ATPases. The resulting $\varDelta\bar{\mu}_{H+}$ is thought to be consumed in driving secondary solute transports H^+ via coupled H^+/solute fluxes.

Further Reading

Epstein, E. 1976. Kinetics of Ion Transport and the Carrier Concept. *Encyclopedia of Plant Physiology, New Series IIB*, eds. U. Lüttge and M.G. Pitman, pp. 70-94. Berlin: Springer-Verlag.

Clarkson, D.T. 1974. *Ion Transport and Cell Structure in Plants.* London: McGraw Hill.

Hodges, T.K. 1973. Ion absorption by plant roots. *Advances in Agronomy* 25: 163-207.

Lüttge, U., and Higinbotham. 1979. *Transport in Plants*. New York: Springer-Verlag.

Nicholls, D.G. 1982. *Bioenergetics. An introduction to chemiosmotic theory*. London: Academic Press.

Poole, R.J. 1978. Energy coupling for membrane transport. *Annual Review of Plant Physiology* 29: 437-60.

Raven, J.A., and Smith, F.A. 1974. Significance of hydrogen ion transport in plant cells. *Canadian Journal of Botany* 52: 1035-48.

Spanswick, R.M. 1981. Electrogenic ion pumps. *Annual Review of Plant Physiology* 32: 267-289.

Sze, H. 1984. H^+-translocating ATPases of the plasma membrane and tonoplast of plant cells. *Physiologia Plantarum* 61:683-691.

6

Long-Distance Transport of Ions

In Chapter 5 the processes responsible for the transport of inorganic ions across plant membranes, particularly the plasma membrane, were discussed in some detail. Having traversed the plasma membrane, the absorbed ions enter the cytoplasmic phase of the cell and in unicellular organisms (for example green algae, such as *Chlorella* or *Chlamydomonas*) will probably travel but a few μm before becoming assimilated or stored for future use.

By contrast, in higher plants the site of inorganic ion absorption (cells of the root) may be separated by distances of up to a hundred metres from the site of major assimilatory activity (the leaves). This is not to suggest that roots are matabolically inert; quite the contrary. However, there is generally much greater metabolic activity in the aerial portions of the plant and most of the inorganic ions absorbed by roots are transferred without change to the shoot. One manifestation of this greater activity is the larger biomass of leafy tissues compared to roots (shoot to root weight ratios are commonly 3 to 4 in most plants). The spatial separation of root and shoot (the typical land plant morphology) reflects, of course, the localization of the plant's primary resources; water and inorganic ions in soil and sunlight and CO_2 in air. The physical separation of functions which has evolved in land plants, and reaches its extreme expression in forest trees, has been made possible by the evolution of an efficient two-way long-distance transport system between roots and shoots. The xylem carries water and inorganic ions to the shoot while the phloem carries a source of free energy (usually in the form of sucrose) from mature leaves to parts of the plant (roots, developing leaves and fruits) where there is a demand for this resource.

In addition, particularly when plants are grown under conditions of poor inorganic nutrient supply, retranslocation of inorganic nutrients from older to younger tissues, within the phloen, may be significant. This chapter deals with the processes which are responsible for the (long-distance) transport of the inorganic elements.

6.1   Nutrient Translocation to the Shoot

Nutrient translocation to the shoot primarily involves the translocation of inorganic ions. However, some species bring about significant reduction of NO_3^- to organic N within their roots and these species may translocate reduced forms of N in additon to NO_3^- and the other inorganic elements (see Section 6.1d). Legumes, in particular, export their reduced N to shoots in the form of amides such as glutamine and asparagine and other more complex amino acid derivatives such as the ureides.

One of the earliest proposals dealing with the phenomenon of translocation was that of the 19th century botanist Sachs who suggested that ions were simply drawn up from soil as an uninterrupted flow of solution in the transpiration stream. Today there are many reasons to reject such a simple explanation, although, as we shall see, it is correct in one minor respect. The processes involved in the translocation of ions to the shoot are numerous and they demonstrate complex interactions. In order to clarify these processes it is convenient to divide translocation into its several component parts; namely ion absorption, transport across the cortex, release to the xylem vessels, and transport to the shoot.

6.1a   ION ABSORPTION

In studying the translocation of ions to the shoot, experimenters have made frequent use of decapitated plants. If the shoot is severed from the root close to its base the root will exude copious quantities of xylem sap from its cut surface under appropriate circumstances. If the cut surface of the root is sealed into a glass tube, the exudate can be collected at intervals and analyzed for ion content. This arrangement provides an ideal system for investigating the characteristics of ion translocation because xylem exudation can be sustained for several hours. When this exudate is analyzed it is characteristically found to be considerably more concentrated than the solution surrounding the roots. Table 6.1 provides data for the concentrations of ions in the external solution and in the xylem exudate of decapitated corn roots collected by Davis and Higinbotham (1969). In the first experiment using 0.1 mM K^+ the exudate was 147 times more concentrated than the external medium. From observations of this sort it should be evident that a simple flow of solution into the xylem as proposed by Sachs could not possibly result in such a concentration of ions.

We have seen (Chapter 4) that the endodermis represents a formidable barrier to the free (apoplasmic) passage of inorganic ions between the cortex and the stele. Consequently it is generally held that before ions can enter the stele they must be absorbed into the cortical symplasm. Many aspects of this absorption have already been dealt with in Chapters 4 and 5. In particular the relative importance of ion uptake by epidermal and cortical cells, and the variation of ion uptake along the root axis were considered. In addition, the development of the endodermis was described with reference to its role as a barrier to apoplasmic transport between cortical cells and the stele. The detailed treatment of ion absorption in Chapter 5 was restricted to transport across plasma membranes, without reference to the subsequent process of long-distance transport. However it must be emphasized again that only about 20 percent of the ions removed from soil, through the action of the roots, remain in these organs; the other 80 percent is translocated to the shoot. The root is therefore

Table 6.1.   Ion concentrations of exudate collected from excised corn (*Zea mays*) roots. (Data of Davis and Higinbotham, *Plant Physiology*, 44:1383-1392, 1969)

	ION CONCENTRATION (mM)					
	K^+	Na^+	Ca^{2+}	Mg_2^+	Cl^-	NO_3^-
External Solution	0.1	0.1	0.1	—	—	0.2
Exudate (from xylem)	14.7	0.8	1.4	—	—	5.6
External Solution	1.0	1.0	1.0	0.25	1.0	1.0
Exudate (from xylem)	19.4	1.2	3.3	1.4	3.1	9.7

much more than an organ for the absorption of ions; it is, in addition, a secretory organ responsible for the delivery of inorganic nutrients to the stele for onward transport. For the purposes of simplifying these functions to their most basic representation, the root can be depicted as a series of subcompartments with ion fluxes to and from the various compartments, as in Figure 6.1, reproduced from the review by Pitman (1977). In this representation the net flux into the root ($\emptyset$) results from the balance between influx ($\emptyset_{oc}$) and efflux ($\emptyset_{co}$). Generally $\emptyset_{co}$ is small compared to $\emptyset_{oc}$ (otherwise there would be no absorption) but at very low external concentrations influx may be so reduced that $\emptyset_{oc} = \emptyset_{co}$ and hence $\emptyset = 0$. This situation is analogous to light or CO_2 compensation in photosynthesis. Moreover, studies of NO_3^- absorption have revealed that during the night $\emptyset_{oc}$ may decline below compensation in some plants so that $\emptyset$ is negative, i.e., there is a net loss from the roots.

Once across the plasma membrane and into the symplasm several

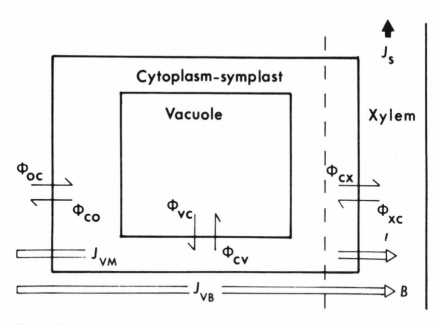

Figure 6.1.   Diagrammatic representation of fluxes between various compartments of the root. Ion fluxes to and from the cortical symplasm ($\emptyset_{oc}$ and $\emptyset_{co}$); and to the xylem: $\emptyset_{cx}$. J_{vm} and J_{vb} represent water fluxes to the xylem via the symplasm and the bypass, respectively. J_s represents the total solute flux to the xylem. (From Pitman, *Annual Review of Plant Physiology*, 28:71-88, 1977)

potential fates (besides efflux) await the absorbed ions. They may enter various membrane-bound organelles; mitochondria, for example, are thought to sequester Ca^{2+} as well as using Pi in the generation of ATP. Other organelles include the nucleus, endoplasmic reticulum (e.r.) and last, but not least, the vacuole. Transport across the tonoplast ($\emptyset_{cv}$) and into the vacuole provides for the generation of turgor as well as forming a large reservoir for nutrient reserves. Figure 6.1 indicates that ions may also leave the vacuole ($\emptyset_{vc}$) and, particularly when external supplies are inadequate, these reserves are drawn upon so that a net loss of ions from the vacuole ($\emptyset_{vc} > \emptyset_{cv}$) results. Because of the inherent difficulty in analyzing fluxes to and from the smaller organelles, *in vivo*, there is inadequate information available to complete this representation.

6.1b TRANSPORT ACROSS THE CORTEX

In terms of the large proportion of absorbed ions which are eventually delivered to the shoot it should be clear that the major proportion of absorbed ions is transported across the cortex to the stele for onward translocation (within xylem vessels) to the shoot. This can readily be demonstrated in seedling plants grown by labeling the external medium with ^{32}Pi or ^{42}K and subsequently analyzing shoot tissue for radioactivity. However, when the ion content of seedling roots is reduced by growth without adequate external reserves (e.g. in dilute $CaSO_4$ solution), producing the so-called "low-salt" roots, the higher priority normally afforded translocation appears to take second place to the filling of vacuoles. Thus, when K^+ is provided to such low-salt roots there is no increase of shoot K^+ content for several hours although root content is continuously increasing during this time. In my laboratory it was observed that transport of K^+ to the shoots of low-salt barley plants began after about 6 h, when root $[K^+]$ had reached its maximum value $\sim60\mu$mol g^{-1} (Fig. 6.2). The control of translocation under these conditions represents an intriguing problem. It might be hypothesized that the three compartments in the root (cytoplasm, vacuole and translocation path) shown in Figure 6.1 are all in competition for absorbed ions. Highest priority naturally goes to the cytoplasm. In low-salt roots it is thought that cytoplasmic ion concentrations are maintained at the expense of the vacuole. When ions become available the latter organelle takes the "lion's share" of incoming ions until its concentration is restored. Subsequently, absorbed ions are mainly destined for the shoot. However, this

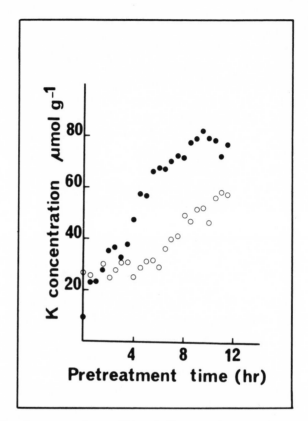

Figure 6.2. Increase of root (•) and shoot (○) K concentration in barley when 6-day old plants grown in CaSO$_4$ solution are pretreated with KCl solution. (From Glass, *Canadian Journal of Botany*, 56:1759-1764, 1978)

"competition" model is not entirely adequate, because, when translocation to the shoot is underway, transfer of plants back to CaSO$_4$ solution can result in continued translocation of K$^+$ to the shoot for many hours until the root vacuoles are back to their original low-salt condition. It has been suggested that the delay in translocation observed when low-salt roots are first exposed to a source of ions may be due to the time required to "induce" the mechanism(s) responsible for translocation. Implicit in this hypothesis is the idea that the absorbed ion is responsible for inducing its own translocation. Some experimental work has been completed in this area but a clear resolution of this question has not yet been provided.

Patterns of ion translocation are somewhat different from patterns of uptake. Vascular differentiation is not apparent until about 0.5 to 1 mm back from the apical meristem, where phloem differentiation precedes that of the xylem. Understandably, the capacity for translocation depends upon the state of differentiation of the root's conducting elements. In barley, for example, the maximum rates of translocation (expressed as percent of that absorbed) do not occur until a few cm behind the root tip. However, despite the extensive depositions of suberin and cellulose in State III endodermal cells at some distance (up to 40 cm) from the tip, the passage of K^+ and Pi through these cells and into the stele occurs as readily as at 6 cm from the root tip where only the Casparian strips are present. The ability to traverse the thickened endodermal wall is thought to depend upon the numerous plasmodesmata (as many as 10^4 per cell) which traverse adjacent endodermal and pericycle cell walls, particularly in localized areas (pits) which remain unthickened (see Fig. 6.3).

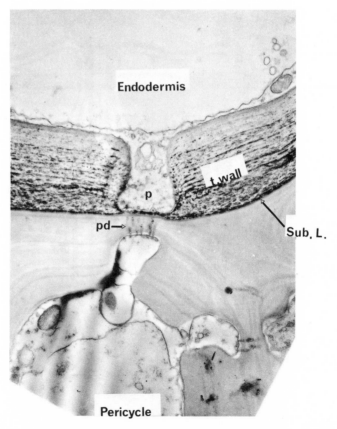

Figure 6.3. Plasmodesmata (pd) linking the cytoplasm of a State III endodermal cell of barley root and the adjacent pericycle cell. Note the massive tertiary wall development and the suberin lamella (Sub. L.). Plasmodesmata in this electromicrograph are localized in the unthickened regions of adjacent cell walls (referred to as a pit pair, p). (By courtesy of Dr. A. W. Robards)

PLASMODESMATA AND SYMPLASMIC TRANSPORT

Ultrastructural studies reveal that plasmodesmata are particularly numerous (as many as 10^6 per mm^2) in cells that are actively involved in cell-to-cell transport of solutes. Each plasmodesma consists of a pore which traverses the cell wall between adjacent cells. The pore is lined by plasma membrane which is continuous from cell to cell. The central cavity of the plasmodesma is occupied by a tubule called the desmotubule, within which is located a central rod. The desmotubule is thought to consist of, and be continuous with, the endoplasmic reticulum (e.r.) of adjacent cells. Figure 6.4 provides a

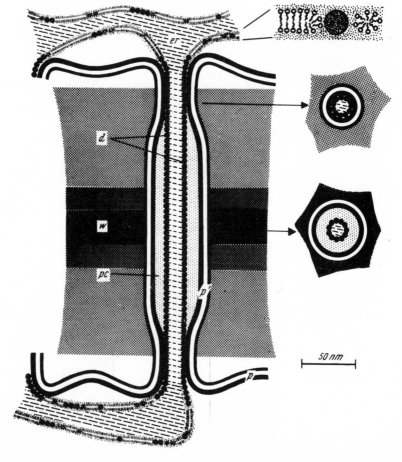

Figure 6.4. Diagrammatic representation of the ultrastructure of a plasmodesma, showing endoplasmic reticulum (e.r.) continuous with the desmotubule (dt). Plasma membrane (p) lines the cavity (pc) of the plasmodesma. At the right are shown plasmodesmata as they appear in a cross section. (From Robards, *Protoplasma*, 72:315-323, 1971)

diagrammatic representation of a plasmodesma connecting two cells according to the interpretation proposed by Robards in 1971. Thus, the structural basis for cellular continuity is provided.

The apparent association between e.r. and the desmotubule has led to the belief that e.r. may be of particular importance in symplasmic transport of solutes. There is also physiological evidence for this idea, based upon the localization of Cl^- (by precipitation as the insoluble AgCl following $AgNO_3$ treatment), in cells involved in cell-to-cell transport of this ion. This precipitation technique has been used successfully in the preparation of tissue samples for electron microscopy. The prominent concentration of AgCl within plasmodesmata and e.r. gives additional evidence of their participation in symplasmic transport.

There is also electrical evidence that the symplasm represents a relatively low resistance pathway for solute transport. By injecting an electrical current into a cortical cell by means of a glass microelectrode while monitoring changes in the electrical potentials of adjacent cells, Spanswick (1972, 1976) was able to estimate the electrical resistance of the plasma membrane, tonoplast and plasmodesmata. His data suggest that plasmodesmata are about 100 times more conductive than the plasma membrane.

It has often been proposed that symplasmic transport would be assisted by the action of cytoplasmic streaming. Indeed, in the giant-celled alga *Chara* the rate of cell-to-cell transport of Cl^- has been shown to be a function of the rate of cytoplasmic streaming. However, in the smaller cells of higher plants, in particular cortical cells of the root, it has been argued that diffusion should be adequate to propel solutes from cell to cell. Over such dimensions ($\sim 20 \ \mu M$) the plasmodesmatal connections rather than rates of intracellular transport should limit the rate of ion movement across the cortex. Using Cytochalasin B to inhibit cytoplasmic streaming, it has been established in my laboratory that rates of K^+ transport across the roots of tomato and cucumber plants are unaffected when streaming is completely terminated.

In contrast to the situation for K^+ and Pi, entry of Ca^{2+} into the stele is limited to the region close to the root tip. In Section 4.4b it was proposed that ion absorption in soil-grown plants is probably restricted mainly to the epidermis and outer cortical cells, and that transport across the cortex must therefore involve symplasmic fluxes. However, where external concentrations of ions are high and/or where uptake rates are low (e.g. when plant demand for a particular nutrient is small) there may be a significant apoplasmic movement across the cortex. This is thought to be the case for calcium. Delivery of this ion up to the root surface by bulk flow may be sufficiently in excess of plant requirement that diffusion gradients actually develop from the root surface to soil. This is exactly opposite to the situation described for K^+ and Pi in Section 3.5. Unlike K^+ and Pi, Ca^{2+} entry to the stele and subsequent translocation declines quite abruptly about 6-10 cm from the root tip in the region corresponding to the development of the suberin lamellae (State II condition of endodermal cells) in barley and marrow roots (Fig. 6.5). Thus it would appear that the suberin lamellae represent a substantial barrier to the movement of Ca^{2+} into the stele. Clarkson and his coworkers have proposed that Ca^{2+} traverses the cortex in the apoplasm and reaches the stele via the

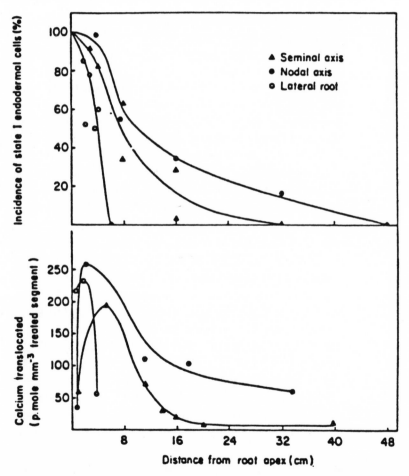

Figure 6.5. A comparison of the distribution of unsuberized (State I) endodermal cells in seminal, nodal and lateral roots of barley and Ca^{2+} translocation from 3.5 mm segments of the intact roots. (From Robards et al. *Protoplasma*, 77:291-311, 1973)

symplasm of State I endodermal cells, after transfer across the plasma membranes of these cells. With the development of the suberin lamellae this entry point is barred. If this explanation is correct it is intriguing that Ca^{2+} is apparently unable to enter into State II endodermal cells via the symplasm from adjacent cortical cells. Perhaps significant symplasmic transport of this ion is simply not possible because of the very low concentration of cytoplasmic Ca^{2+} (see Chapter 8).

Whether ion fluxes are completely symplasmic or apoplasmic up to the endodermis, it is evident from the above discussion that most ions must eventually enter the endodermal symplasm. However, it is thought that there may be a small additional solute flux which enters the stele via the apoplasm, completely bypassing the endodermal cytoplasm. The exact location of this "straight-through" flux is uncertain but the sites of lateral root emergence may be responsible. As lateral root primordia grow out across the cortex (Section 4.3) the endodermal barrier may be broken, providing an uninterrupted apoplasmic pathway to the stele. Only a very small percent ($\sim 1\%$) of the transpiration stream enters the stele in this way so that under most conditions the amount of solute entering the stele by this route is insignificant. However, when external ion concentrations are high this flux may assume greater importance. Pitman (1977) has estimated that in 50 mM NaCl solution this uncontrolled flux might represent 10-20% of the observed translocation rate. For plants such as mangroves and other halophytes whose roots are bathed in sea water (500 mM NaCl) such a flux could represent a major problem.

6.1c RELEASE OF IONS TO THE XYLEM

One of the earliest mechanisms proposed for the release of inorganic ions to the lumens of xylem vessels was that of Crafts and Broyer (1938). Their model was based upon the hypothesis that ions accumulated within the symplasm are released to stelar xylem vessels as a result of passive leakage. A low O_2 tension at the centre of the root was advanced as a possible cause of the "leaky" plasma membranes. However, measurements of the O_2 tension across the cortex and into the stele by means of oxygen microelectrodes have revealed only a small gradient of this gas.

Since the original hypothesis by Crafts and Broyer, two schools of thought have emerged on this topic, which are sometimes referred to as the "one pump" and "two pump" hypotheses. According to the one pump hypothesis, which is essentially in agreement with the Crafts/Broyer model, ions are actively "pumped" into the symplasm but move passively from the root symplasm into the xylem vessels. Bowling and his coworkers have used ion-selective microelectrodes and electropotential electrodes to measure the electrochemical potentials across the root. This group has claimed that fluxes of K^+, Cl^-, SO_4^{2-} and NO_3^- from symplasm to xylem vessels do not involve movement against electrochemical potential gradients; i.e. they are passive fluxes.

According to supporters of the two-pump hypothesis, ions are actively secreted into the symplasm and entry into xylem vessels involves a second active transport step. Advocates of this model have criticized the findings of Bowling and have published data which appear to show that K^+ moves against its $\overline{\Delta\mu}$ into the xylem. Further evidence in support of this model is given by the following observations:

(a) Metabolic inhibitors frequently cause a rapid cessation of ion transport to the xylem. Although it can be argued that inhibitors might reduce ion uptake and thus affect transport to the stele indirectly, it is known that ion transport can occur for many hours after an external source of ions has been removed, presumably by withdrawal of ions stored in cortical vacuoles. Thus, inhibition of uptake alone should not inhibit transport so readily unless a metabolically dependent step is involved in translocation. However, transport of K^+ from vacuole to cytoplasm could represent this active transport step.

(b) Pitman, formerly at the University of Sydney, Australia, has claimed that certain inhibitors e.g. parafluorophenylalanine (an analog of phenylalanine which results in the synthesis of nonfunctional proteins) and the growth regulator abscisic acid appear to exert a strong effect upon translocation while producing only a minor or delayed effect upon ion uptake.

(c) Using the technique of x-ray microprobe analysis to localize elemental distribution across roots of barley, Läuchli, at the University of California, demonstrated that xylem vessels and xylem parenchyma cells contained higher levels of K^+ than did cortical cells, thus raising doubts about a passive gradient from cortex to stele.

If the radial transport of inorganic ions does involve two active transport steps, which cells are involved in the second active transport? An early proposal by Hylmo (1953) was that the release of ions into the lumens of xylem vessels was through the tonoplasts of partially differentiated xylem vessels. This has been called Hylmo's test-tube hypothesis and has been supported in more recent years by various other groups. In essence the model is based upon the idea that partially differentiated vessels could be envisaged as having intact membranes (plasma membranes and tonoplasts) at their lower (younger) ends while at their upper differentiating region these membranes would be lost. Thus ions could proceed within the symplasm to the young portion of the vessel and be transported into the vacuole of the vessel. At its upper end the vacuole would be continuous with the cavity (lumen) of the vessel (see Fig. 6.6).

However, published reports showing that ions can traverse the cortex and enter xylem vessels at considerable distances from the root tip (where all xylem differentiation has long since been completed) make it difficult to sustain this hypothesis. As an alternative to the test-tube hypothesis, there is now considerable support for the idea that xylem parenchyma cells act as the site for loading ions into the xylem vessels. This is based upon the following considerations:

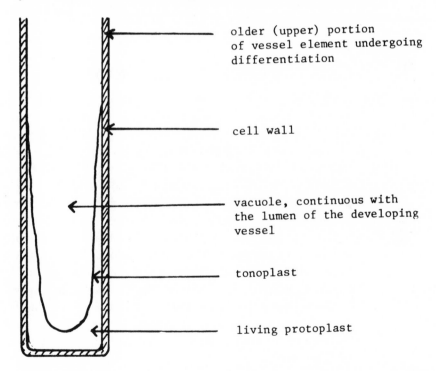

older (upper) portion
of vessel element undergoing
differentiation

cell wall

vacuole, continuous with
the lumen of the developing
vessel

tonoplast

living protoplast

Figure 6.6.   A diagrammatic representation of the Hylmo "test-tube" hypothesis for delivery of ions to the lumen of xylem vessels. Note that the vacuole of the protoplast in the lower (living) portion of the protoplast is continuous with the lumen of the upper (older) differentiating portion of the vessel.

(a) Location: xylem parenchyma cells are located adjacent to the outer (metaxylem) vessels which are mainly responsible for the upward translocation of inorganic ions.

(b) Cytology: xylem parenchyma cells possess dense cytoplasm, with well developed membrane systems, limited vacuolation and numerous mitochondria. These are all good indications of a metabolically active cell type.

(c) Ion Content: localization of ions by autoradiography, X-ray microprobe analysis, and precipitation of insoluble salts (e.g. AgCl) indicate that administered ions are particularly concentrated in xylem parenchyma and outer metaxylem vessels.

It is clear from the current literature that the question of active secretion of ions into the xylem, as opposed to passive (though possibly carrier-mediated) transfer, is not yet resolved. There are shortcomings in the arguments and in the data advanced by both groups. As much as we would prefer a neat resolution of the problem we are forced to accept that an unequivocal answer to the question of one versus two pumps is not yet forthcoming.

6.1d TRANSPORT TO THE SHOOT

Once inorganic solutes enter into the xylem vessels they are transferred to the shoot by bulk flow within the transpiration stream. The driving force for this convective flux is purely passive; the difference of water potential ($\Delta\Psi$) between xylem sap and air. Typically, when xylem sap is collected as exudation from the cut surface of a decapitated plant, it will contain about 1 - 20 mg of dry matter ml^{-1}. The pH of this sap tends to be slightly acid (5.2 - 6.5) compared to the phloem sap (~ 8.0) and is generally quite a dilute solution compared to that contained in the phloem (see Table 6.3). Generally the xylem sap contains the essential elements such as K^+, Ca^{2+}, Mg^{2+}, Pi, Cl^-, NO_3^- and SO_4^{2-} in the form of the inorganic ions but, particularly where significant reduction of NO_3^- occurs in the roots, and in legumes where N_2 reduction occurs within root nodules, various reduced forms of N, may be present in significant quantity. In addition Fe may be transported in xylem fluid chelated with organic acids such as citrate.

En route to the shoot, ions may be unloaded to various tissues and organs. Moreover, there is strong evidence that potentially toxic elements such as Na may be removed en route so that the concentration of this element reaching the shoot is much reduced. It appears that the flow of divalent cations (Ca^{2+} and Mg^{2+}) may interact with the walls of the vessel element which behave like a cation exchanger, thus slowing their transport.

6.1e TRANSPIRATION AND ION TRANSLOCATION

The picture generated thus far is that inorganic solutes enter xylem vessels and are then transported passively along the vessels in response to the transpiration-driven water flux. However there is considerable evidence to show that both uptake and translocation rates of several ions can be influenced by the velocity of transpiration. Broyer and Hoagland (1943) established that rates of ion uptake by roots of intact barley plants could be increased in response to increased rates of transpiration. In soil-grown plants it might be anticipated that the delivery of solutes up to and into the root cortex by bulk flow might be increased under conditions of high transpiration. Similar effects might well operate under experimental conditions using hydroponically-grown roots. Rates of ion transport to the xylem have repeatedly been shown to respond to elevated rates of transpiration. The data of Table 6.2 collected by Greenway (1965) demonstrate that

Table 6.2. Interactions between Transpiration and Cl^- Translocation in barley. (Taken from Greenway, *Australian Journal of Biological Sciences*, 18:249-268, 1965)

External Solution	A Transpirational Water Flow $(mm^3 m^{-2} s^{-1})$	B Total translocation of Cl^- to the Shoot $(nmol\ m^{-2} s^{-1})$	C (B/A) Cl^- Concentration of Xylem Sap (mM)
50 mM NaCl plus	2.2	34	15.5
nutrient solution	2.2	31	14
	4.1	41	10
	5.0	40	8
	8.7	70	8
	13.0	65	5
	14.5	43	3
	15.0	60	4
	18.0	72	4
	19.0	76	4
	35.0	70	2

total Cl⁻ translocated to barley shoots was almost doubled as transpiration increased from 2.2 to 35.0 $mm^3m^{-2}s^{-1}$. It is evident that as the flow of xylem sap increased, due to increased transpiration, the xylem sap was diluted from 15.5 to 2mM. However, whereas transirtion was increased almost 16-fold the xylem sap was only diluted by a little more that 8-fold. The overall effect was a greater net transfer of Cl⁻ to the shoot. Essentially similar observations have been reported for several other ions, e.g. Na^+, Ca^{2+}, Mg^{2+}, SO_4^{2+} and Pi. This link with transpiration may be due to the endodermal "bypass" referred to earlier. Particularly at the concentration employed in the experiment by Greenway (50mM NaCl), such an effect would be anticipated. In the case of Pi translocation, Scott Russell (1977) has reported that high rates of transpiration caused increased rates of Pi translocation when the external solution was maintained at 1 mM Pi (which is considerably higher that soil solution strength) but observed no such effect at $0.03\mu M$. Clearly the "bypass" explanation is consistent with these observations. However, the "flushing" of solutes from xylem vessels during high rates of transporation may facilitate greater solute transport into the stele by reducing the gradient for the putative second active transport step, or by altering the permeability properties of the critical membranes.

6.2   Retranslocation

In Section 6.1a it was noted that approximately 80 per cent of all ions absorbed by plant roots are translocated to shoots. Under conditions of abundant nutrient supply, as for example in hydroponic facilities, ions absorbed by the roots are delivered directly to growing regions (particularly shoots) via the xylem translocation pathway. Moreover, absorbed ions may appear in leaf tissue within 30 to 60 minutes of their removal from solution by the root system.

However, under field conditions soil solutions may become so depleted at particular times of year that potential growth rates can no longer be sustained from external resources. This situation may be true for wild plants in nutrient-poor soils, but can also apply in the agricultural context, where fertilizers, applied at planting time, are severely depleted later in the season. At a time which could be critical for crop productivity, exogenous supplies of nutrients may therefore be quite inadequate to satisfy plant demands.

Plants appear to have solved this potential problem by accumulating inorganic nutrients in excess of their immediate requirements.

The reserves which are thereby created can be drawn upon subsequently, when external supplies become limiting. Cereal plants, such as wheat and oats, for example, typically accumulate over 90 per cent of their total N and P content before plants have attained 25 per cent of their final dry weight. Particularly during the reproductive phase of the life cycle, as much as 70 to 90 per cent of P, N, K and Mg may be translocated from leaves to satisfy the considerable demands of developing fruits and seeds.

The subcellular location of these reserves (in both roots and leaves) is the vacuole. Examination of a typical parenchyma cell (e.g. a mesophyll cell from the leaf of ryegrass: Fig. 6.7) reveals that the vacuole occupies greater than 90 per cent of cell volume. Using vacuoles isolated from barley leaf protoplasts, it has been estimated that ~90% of cellular NO_3^- is within vacuoles. The combined vacuolar volume of all such cells, therefore provides a considerable storage capacity.

The use of radioisotopes has revealed that the redistribution of elements from mature tissue to rapidly growing younger plant parts occurs even when the plant is not under nutrient stress. However, when such conditions do apply, retranslocation assumes considerably greater importance. Also, under harsh environmental conditions such

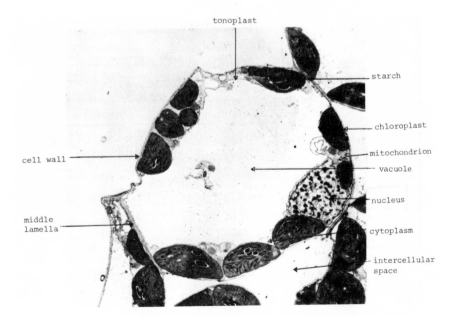

Figure 6.7. An electron micrograph of a mesophyll cell from ryegrass (magnification 7500). (Courtesy of the Department of Scientific and Industrial Research, Palmerston North, New Zealand)

as extremes of cold (in Arctic plants) or heat (in desert plants) nutrients may be translocated from above ground tissues prior to their death, to be stored in below ground organs until conditions favourable to growth return.

The process of redistribution of absorbed nutrients, therefore, appears to be universal among multicellular plants. However, certain crop plants, notably cereals such as wheat and oats, cotton and tobacco, and many of the legumes, have been the subject of intensive study.

By making incisions into the fruits of a variety of legume species, Pate and his colleagues at Western Australia have been able to collect the sap which "bleeds" from the severed conducting tissue. Table 6.3 lists the major components of the sap entering fruits of lupin (*Lupinus alba*).

In perennial species such as forest trees retranslocation of nutrients is not only extremely important for the survival of the individual tree, but has critical implications for the stability of the whole forest ecosystem. In both deciduous and evergreen trees, there is substantial withdrawal of N, P, K and S from leaves prior to leaf fall. This source of nutrients is stored within the permanent tissues, and may provide a substantial proportion of the requirements for new growth. In a

Table 6.3. A comparison of the solute content of xylem and phloem sap exuding from the cut surface of lupin (Lupinus albus) fruits. (Data from Pate, *Encyclopedia of Plant Physiology*, 1:451-473, M.H. Zimmerman and J.A. Milburn, eds. Springer-Verlag, (1975)

Solute	*Xylem Sap*	*Phloen Sap*
sucrose	not detected	154 mg ml^{-1}
amino acids	0.70 mg ml^{-1}	13 mg ml^{-1}
nitrate	0.16 mM	not detected
potassium	2.3 mM	39.5 mM
sodium	2.6 mM	5.2 mM
magnesium	1.1 mM	3.5 mM
calcium	0.4 mM	0.5 mM
iron	0.03 mM	0.2 mM
manganese	6 μM	14.6 μM
zinc	6 μM	88 μM
copper	trace	6 μM

49-year-old Douglas fir stand, e.g. it was estimated that 40% of N, 50% of P and 14% of K required for new growth came from such reserves. Were these nutrients to be lost from the tree during leaf fall there is a possibility that they might be permanently lost as a result of leaching from the soil. This possibility is particularly significant in tropical forests. During defoliation programs in the Viet Nam War for instance, considerable ecological damage resulted when leaves were caused to drop prematurely (as a result of spraying with herbicides) before critical inorganic nutrients could be withdrawn. Interestingly, in species of *Alnus* (alder), which establish symbiotic associations with nitrogen fixing fungi (Chapter 8) and where N is presumably not in short supply, leaves released in Autumn still contain a large amount of N. This is one means whereby *Alnus* increases the fertility of soil supporting this species.

Among plant nutritionists it is a well known adage that the location of the earliest visible deficiency symptoms (in young as opposed to older tissues) provides a useful guide to the identity of the deficiency. This arises because certain nutrients are particularly mobile in the plant and readily retranslocated from older to younger leaves when external supplies of the element are inadequate. Good examples of such mobile elements are N, P, K and Cl. As a consequence of their removal from older leaves their associated deficiency symptoms first appear in these tissues. Certain other elements, notably Ca and B are relatively immobile; they tend to be retained by older leaves and, as a result, deficiency symptoms show up first in the younger leaves.

Retranslocation is considered to occur within the phloem. The use of the aphid-stylet technique, in which aphids are allowed to draw phloem sap directly from sieve tubes by means of their long mouth parts, has provided a convenient method for analyzing the contents of sieve tube sap. Typically the sap is rich in K, P, and Mg (though this may not necessarily be related to retranslocation). Nitrogen and sulfur are present in organic form, while Ca^{2+} is present at quite low levels.

In the literature dealing with phloem translocation of photosynthate the terms "source" and "sink" are used to designate regions where carbohydrate is synthesized (mature leaves) and consumed (expanding leaves, roots, fruits etc.), respectively. Likewise an older leaf may represent the source for translocation of K or P toward a younger leaf (the sink). The translocation of solutes (including sugars) from source to sink has become a well established concept for phloem transport. However, the underlying mechanisms responsible for this directional translocation have been the subject of intense debate.

Although it is certainly not universally accepted, the Münch Pressure Flow Hypothesis of phloem transport of sugars appears to have the support of a majority of scientists. The mechanisms responsible for retranslocation of the inorganic elements is even less clearly understood.

Numerous experimenters have sought to link plant growth regulators with patterns of translocation within phloem. For example, cytokinins can inhibit the mobilization of nutrients from mature leaves. This may be the cause for their well known capacity to retard senescence of mature tissue. Wareing and his collaborators at Aberystwyth in Wales have shown that auxins, cytokinins and gibberellins are all capable of intensifying the tendency for solutes (e.g. ^{32}Pi) to accumulate at the sink. A much used experimental tool is the decapitated seedling or debudded twig. When natural sinks (e.g. expanding leaves or buds) have been removed, applications of the above growth regulators to the stem apex have been shown to cause increased rates of transport (within the phloem) to the apex. It is therefore reasonable to presume that centres of rapid growth, which represent rich sources of the growth regulators, may exert a potent influence upon translocation through the action of the latter compounds. This is certainly far from a complete explanation of the control of translocation. However, as in many other areas of plant physiology, our understanding of the details of this process is still quite superficial.

Summary

Most of the inorganic ions absorbed by plant roots are translocated in the transpiration stream to the shoots. Under steady state conditions recently absorbed ions pass across the cortex and enter the stele via the symplasm. However, under "starved" conditions it is several hours before translocation to the shoot can be detected. During this period ions appear to be transported across the tonoplasts into the vacuoles of depleted cortical cells.

Entry to the stele is almost exclusively via the symplasm of the root, the apoplasmic pathway being obstructed by the specialized thickenings of endodermal cells, the Casparian strips. Even when secondary and tertiary wall development occurs in these cells the plasmodesmatal connections maintain continuity for symplasmic transport to the stele. Ions such as K^+ and Pi can readily pass through the State II and State III endodermal cells. However, this is not the case for Ca^{2+} which can enter the stele only via State I endodermal cells.

Release of ions to the xylem vessels may involve an active transport step from the symplasm which may be mediated by xylem parenchyma cells. Retranslocation of ions (within the phloen), from older to younger tissues, appears to be widespread among plants. Under harsh environmental conditions, including nutrient stress or prior to leaf abscission this process appears to take on even greater importance.

Further Reading

Anderson, W.P. 1976. Transport through roots. In *Encyclopedia of Plant Physiology, New Series, Vol IIB*, eds. U. Lüttge, and M.G. Pitman, pp. 129-56. Berlin: Springer-Verlag.

Bowling, D.J.F. 1976. *Uptake of Ions by Plant Roots*. London: Chapman and Hall.

Bowling, D.J.F. 1981. Release of ions to the xylem in roots. *Physiologia Plantarum*, 53:392-97.

Läuchli, A. 1972. Translocation of inorganic solutes. *Annual Review of Plant Physiology*, 23:197-218.

Läuchli, A. 1976. Apoplasmic transport in tissues. In *Encyclopedia of Plant Physiology, New Series, Vol IIB*, eds, U. Lüttge and M.G. Pitman, pp. 3-34.

Läuchli, A., Pitman, M.G., Lüttge, U., Kramer, D., and Ball, E. 1978. Are developing xylem vessels the site of ion exudation from root to shoot? *Plant, Cell and Environment*, 1:217-23.

Pitman, M.G. 1977. Ion transport into the xylem. *Annual Review of Plant Physiology*, 28:71-88.

Pitman, M.G. 1982. Transport across plant roots. *Quarterly Review of Biophysics*, 15:481-554.

"Transport systems simply cannot be viewed as isolated self-serving mechanisms. Rather, they must be viewed as the first stages in a complex series of processes which constitute plant nutrition."
GLASS, 1987.

7

Environmental Influences on Ion Absorption

Ion transport across cell membranes has frequently been treated as though it were a discrete process, unconnected (apart from its energy requirements) to other metabolic processes. This bias is partly due to the present day tendency toward specialization among researchers who are forced, by the complexity of natural systems, to isolate a part of the system for intensive study. No criticism of this approach is intended, but its inherent dangers have to be acknowledged. The outcome of "isolating" ion transport is that it has frequently been viewed as a necessary (though rather complicated) preliminary to the real business of plant metabolism. A more constructive perspective is to consider the transport of essential elements across the plasma-lemma as a critical first step in metabolism. For example, the assimilation of NO_3^- to yield amino acids, proteins and ultimately to support growth cannot possibly occur more rapidly than the rate of NO_3^- absorption from the external environment. Where NO_3^- is in poor supply, this first step in NO_3^- assimilation may therefore represent the rate limiting step for growth and (particularly for crop plants) yield. However, when supplies of inorganic nutrient are sufficient to satisfy demands imposed by growth the reverse situation may apply, namely, that uptake rates may be limited by the current growth rate. The latter may, in turn, depend upon exogenous factors (such as light or temperature) or endogenous factors (such as relative growth rate) which determine growth. In order to gain a more complete understanding of ion transport processes and their regulation it is essential that the physiologist address the interactions between ion absorptions and the above factors even though the integration of these processes (for the physiologist, not the plant!) represents an extremely difficult task. In addition, scientists with a more ecophysiological bias have long been interested in studying environmental effects upon ion transport with a view to understanding plant competition, and adaptation to environmental stresses.

Experiments designed to investigate the mechanisms of environmental effects on ion transport are usually conducted over relatively short time periods (min to h) so that direct (rather than indirect) effects of the particular variables can be assessed. Typically the experimenter will employ an experimental strategy I refer to as a *perturbation* technique because the plant is perturbed from its original state by the newly imposed variation. This approach has been widely practised in studying ion absorption and has provided valuable insights into the nature of transport processes. It has one serious disadvantage, however, in that it overlooks the possibility that plants may respond to the imposed change (over a time scale of days or weeks) in such a way as to counteract (or reverse) the perturbing influence. As a consequence, long-term predictions which are based upon short-term observations may prove to be erroneous. As an alternative to perturbation experiments it is possible to conduct long-term experiments in which processes of interest, for example rates of ion absorption, are measured over several days under conditions of constant temperature, ion concentration etc. It is essential that experimenters who are interested in environmental effects undertake both long-term investigations and perturbation experiments. This dual approach has not only provided a clearer picture of how environmental variables influence the acquisition of inorganic nutrients but it has also revealed the existence of endogenous regulatory processes which appear to counteract and minimize the effects of environmental perturbations. Our understanding of the way these processes work is in its infancy, but nevertheless these are among the most fascinating and exciting aspects of membrane transport. Moreover, they place the details of membrane transport into a whole organism/ecological context, providing insight into the ways that plants adapt to their environment.

In this chapter some of the more important environmental influences on ion transport are examined, together with the long-term adaptations (acclimations) which serve to regulate tissue composition in the face of these potential perturbations. Excluded from discussion are toxic effects due to extremes of pH, or salinity. These are treated separately in Chapter 9.

7.1   External Concentration

7.1a   PERTURBATION EXPERIMENTS

One of the most important and extensively studied influences on rates of ion absorption by plant tissues is external concentration. You will recall from Section 5.5 that uptake rates at low external concentrations (< 1 mM), which are characteristic of most soils, demonstrate a hyperbolic dependence on external concentration. This was first reported for plant systems by Van den Honert (1937) who demonstrated that phosphate absorption by sugar cane plants conformed to this pattern. Subsequently, many investigators, notably Epstein and his collaborators, have used the concentration-response curve (sometimes called an absorption isotherm) to probe the nature of the absorption process.

It must be appreciated from the outset that such absorption isotherms represent perturbation experiments. Plants are typically grown with inorganic nutrients provided at a particular concentration and then, during a short uptake period, lasting 10-30 min, subsamples of the roots are exposed to different external concentrations. Uptake rates are then calculated on the basis of these short exposures. For example, using 10-day old seedlings of the barley variety Fergus, grown in nutrient media maintained at 5 μM K$^+$, we measured K$^+$ influx (on the basis of 10 min measurements) in the range from 5 to 100 μM K$^+$ (Fig. 7.1). V_{max} and K_m for K$^+$ influx were calculated to be 11.72 μmol g^{-1} h^{-1} and 41 μM, respectively, and the roots used in these experiments were found to contain 32 μmol K$^+$ g^{-1} prior to the influx measurements. Such results are fairly typical of data obtained in many laboratories.

7.1b   LONG-TERM EXPERIMENTS

Figure 7.1 reveals that in the concentration range from 10 to 100 μM K$^+$, influx increased from 2.6 to 8.3 μmol g^{-1}h^{-1} (a three fold increase). On the basis of such short-term (perturbation) experiments it might be reasonable to anticipate that average rates of ion uptake, tissue K$^+$ concentrations and growth rates might respond in similar fashion when plants are grown in this concentration range. However, when the same barley variety was grown for several weeks at various

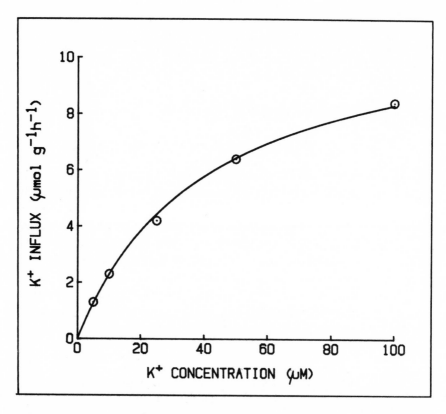

Figure 7.1. K^+ influx as a function of external K^+ concentration into intact roots of Fergus barely plants (*a perturbation experiment*). V_{max} and K_m for influx were calculated to be 11.72 μmol $g^{-1}h^{-1}$ and 41μM, respectively. (From Siddiqi and Glass, *Canadian Journal of Botany*, 61:1551-1558, 1983)

levels of K^+ (5, 10, 50 and 100 μM K^+) which were continuously replenished to maintain constant K^+ concentrations, average rates of K^+ uptake (measured over a two week period), root and shoot K^+ concentrations and even growth rates were essentially independent of external K^+ levels above 10μM. Figure 7.2 shows a plot of K^+ uptake (averaged over the two weeks) for each group of plants. In the same concentration range as that investigated in the perturbation experiment (Fig 7.1), between 10 and 100 μM K^+, uptake increased by only 1.18 (from 1.76 to 2.02 μmol $g^{-1}h^{-1}$) compared to the 3-fold increase seen in the perturbation experiment. The independence of external concentration observed in long-term experiments has consistently been noted by other researchers. For example, Williams (1961) maintained barley plants at ambient K^+ concentrations ranging from 0.25

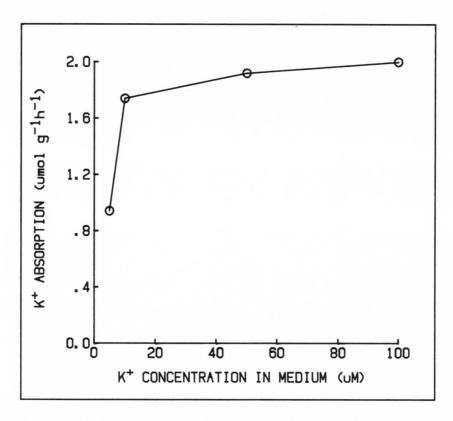

Figure 7.2. Average rates of K^+ absorption by roots of intact Fergus barley plants, measured (over a two week period) at the concentration provided in the growth medium. (Glass, unpublished)

to 1250 μM. Despite this enormous range of concentration, K^+ content of shoots and growth rates increased by factors of only 1.49 and 1.77, respectively. Woodhouse et al. (1978) reported that two weeks after germination, K^+ fluxes were independent of K^+ supply down to 1.25 μM. Although there is more information available regarding K^+, essentially similar observations have been recorded for Pi and NO_3^-.

How can we account for the apparent contradiction implicit in these two sets of data? Clearly, in the perturbed state (i.e. when fluxes are measured at concentrations which are different from those provided during previous growth conditions) influx rates are extremely sensitive to external concentrations. This is undeniable. Figure 7.3 shows several influx isotherms (perturbation experiments) obtained by measuring influx in the concentration range from 5 to 100 μM K^+

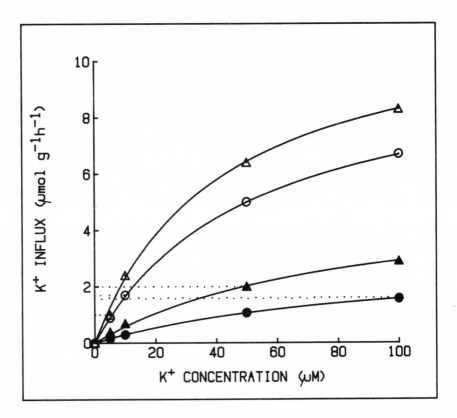

Figure 7.3. K^+ influx as a function of external K^+ concentration (*a perturbation experiment*) into roots of intact Fergus barley plants which had previously been grown at 5 ($\triangle$), 10 ($\bigcirc$), 50 ($\blacktriangle$) and 100 ($\bullet$) μM K^+. The dotted lines from each curve indicate the predicted fluxes at an external concentration corresponding to that which had been provided during the previous growth period. (Glass, unpublished)

in plants previously grown at 5, 10, 50, 100 μM K^+. Three results of this experiment require emphasis:

1. Regardless of the prior growth conditions *all groups of plants retained a strong sensitivity to external concentration* when perturbed. Concentration responses give typical rectangular hyperbolae from which V_{max} and K_m were calculated.

2. *Rates of uptake are strongly influenced by the prior growth regime.* Plants maintained at 5 μM prior to the influx period gave influx values which were greater than 5 times those of plants grown at 100μM K^+. The ability to increase transport capacity in response to low availability of nutrients was first documented by Hoagland and Broyer in 1936. Their observation has been widely

exploited to obtain plants with high rates of uptake by many subsequent investigators. The plants employed for study are routinely grown in a dilute $CaSO_4$ solution for approximately five days before experimentation. Growth under these conditions (at the expense of seed reserves) produces the so-called "low-salt" roots, characterized by high rates of ion uptake, low internal salt levels and high sugar content. This appears to be a universal phenomenon which has been verified for a large number of inorganic ions using a wide range of plant tissues.

Table 7.1 lists V_{max} and K_m values for K^+ influx calculated from the data of Fig 7.3 for the barley variety Fergus. Growth at elevated K^+ levels clearly results in strong reductions of V_{max}, while K_m tends to increase. Decreased V_{max} values have been reported for several inorganic ions. Table 7.2 shows data for

Table 7.1. V_{max} and K_m values for K^+ influx into roots of Fergus barley. Fluxes were measured in short-term (perturbation) experiments using plants previously grown at 5, 10, 50 and 100 μM K^+. (From Siddiqi and Glass, *Canadian Journal of Botany*, 61:1551-1558, 1983)

$[K^+]$ provided in growth medium	V_{max} ($\mu mol\ g^{-1}h^{-1}$)	K_m (μM)
5	11.72	41
10	10.08	51
50	5.22	79
100	2.98	89

Table 7.2. Effect of growth with or without SO_4^{2-}, Pi or Cl^- on the V_{max} and K_m values for absorption of labeled SO_4^{2-}, Pi or Cl^- in a subsequent uptake period. (Data from Lee, *Annals of Botany*, 50:429-449, 1982)

Nutrition During Growth		Ion Uptake Measured	V_{max} ($nmol\ g^{-1}h^{-1}$)	K_m (μM)
A	with SO_4^{2-}	$^{35}SO_4^{2-}$	53.4	13.9
	minus SO_4^{2-}	$^{35}SO_4^{2-}$	758	17.6
B	with Pi	^{32}Pi	257	6.6
	minus Pi	^{32}Pi	475	4.9
C	with Cl^-	$^{36}Cl^-$	1010	57.4
	minus Cl^-	$^{36}Cl^-$	2600	23.7

SO_4^{2-}, Pi and Cl^- uptake in barley roots, reported by Lee at the Agriculture Research Council Laboratories in Letcombe, U.K. In each case one ion was omitted from the growth medium. In a subsequent uptake period K_m and V_{max} values were determined for that particular ion. The data show that withholding a particular nutrient causes substantial increase of V_{max} values.

3. *Even in perturbation experiments, influx measured at the same concentration as the plant has been grown in appears to be essentially independent of external concentrations.* In Figure 7.3 I have used dotted lines to demonstrate influx values for each group of plants (grown at 5, 10, 50 and 100 μM K^+) at the concentration corresponding to that in which they were grown. It is apparent that all except the $5\mu M$ plants gave fluxes close to 2 μmol $g^{-1}h^{-1}$. Note that this is close to the average value given for K^+ uptake in Figure 7.2 for steady state uptake. Almost identical results have been obtained by Deane-Drummond (1981) for NO_3^- uptake by barley. It appears therefore that, given sufficient time to respond, the rates of uptake will approximate the 'required' rate of absorption regardless of external concentration. However, plants cannot work miracles! When the available concentration is too low, as was the case for the plants grown at $5\mu M$ K^+, they are unable to increase their capacity sufficiently to absorb the required 2.0 μmol $g^{-1}h^{-1}$ at 5 μM K^+. These plants showed reduced growth rates and low tissue K^+ levels.

Transport systems simply cannot be viewed as isolated self-serving mechanisms. Rather, they must be viewed as the first stages in a complex series of processes which constitute plant nutrition. As such they are the logical sites for regulatory signals which control their activity in accord with plant demand and nutrient availiability. Ideally, such feedback signals would make the plant independent of external variables such as nutrient availability. Figure 7.4 shows the average growth rates and tissue K^+ concentrations, as a function of available external K^+, for fourteen plant species investigated by Asher and Ozanne (1967). Potassium concentrations required to half saturate growth and tissue K^+ were 5.0 and 7.0 μM K^+, respectively. It is evident that over an extremely wide range of external concentration long-term rates of absorption and plant dry weights were independent of external concentration. However, it should be obvious that the plant's capacity to increase rates of nutrient absorption is not limitless. Beyond some lower level of availability limitation of growth occurs and deficiency symptoms develop.

When plants are grown in nutrient solutions provided with all but

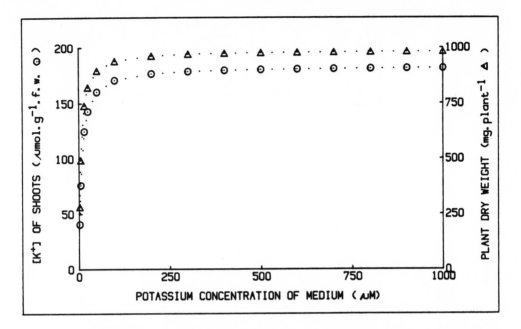

Figure 7.4. Averaged K$^+$ concentrations and dry weights for shoots of 14 species, maintained at the K$^+$ levels provided. (From Asher and Ozanne, *Soil Science*, 103:155-161, 1967)

a single nutrient (for example $-P$ or $-N$) it is generally found that the increased capacity for absorption applies specifically to the nutrient which has been withheld. However there is also evidence for more complicated interactions between ions. Cram, at the University of Sydney, has shown that both Cl$^-$ influx and net NO$_3^-$ uptake are reduced by the prior accumulation of NO$_3^-$ or Cl$^-$ in carrot and barley. As a result, Cl$^-$ influx has been found to be negatively correlated with log (Cl$^-$ + NO$_3^-$) concentration. Such interactions may have important practical consequences.

For example, Dr. Kafkaki, (while at the University of California, Riverside) has shown that in tomato plants the dry weight of tops was strongly depressed and [Cl$^-$] significantly increased as [Cl$^-$] of nutrient media were increased above 10mM (Fig. 7.5a). As external [NO$_3^-$] was increased in these media, the Cl$^-$ content of the plants declined significantly (Fig. 7.5b). These findings may have important consequences for the salinity problem (Chapter 9). Irrigation waters containing [Cl$^-$] above 10mM may cause severe problems for certain crops. However the interactions referred to above indicate that, provided NO$_3^-$ concentrations are maintained at relatively high levels, Cl$^-$ accumulation can be prevented.

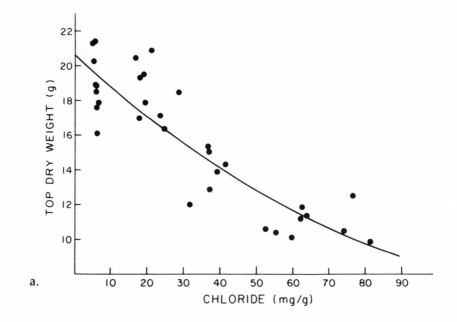

a.

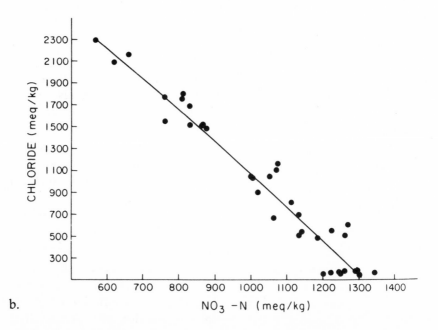

b.

Figure 7.5. (a) Relationships between top dry weight and tissue Cl^- content (mg g^{-1} dry weight) in tomato (from Kafkafi et al. 1982) (b) Relationship between tissue Cl^- and tissue NO_3^- content (meq hg^{-1} dry weight) in tomato. (From Kafkafi et al., *Journal of Plant Nutrition*, 5:1369-1385, 1982)

7.1c MECHANISMS FOR THE REGULATION OF NUTRIENT UPTAKE

It should now be apparent that when nutrients are in short supply plants are capable of adapting to the prevailing conditions by means of both morphological (Chapter 4) and physiological changes. Our understanding of the mechanisms responsible for these adaptations is rudimentary. We are aware of their existence and know something about their properties; there have even been some models proposed to account for these processes. However, considering the limits of our understanding of transport processes *per se*, it is perhaps not surprising that we know so little about the control of their operations.

REGULATION OF UPTAKE: INFLUX OR EFFLUX?

If we accept that the purpose of selective transport systems is to accumulate specific solutes to concentration which are appropriate for the metabolic and osmotic functions they perform (see Chapter 8) then it might be reasonable to postulate that internal ion concentration (or some concentration-dependent variable, such as turgor) would make the ideal feedback signal to regulate rates of ion uptake. As tissue concentrations rise with increasing availability of nutrients, net uptake might be reduced either by a reduction of influx or an increase of efflux. The situation has been found to vary according to the tissue under examination and according to the particular ion investigated or even its concentration range. For rapidly growing plants at low ambient ion concentrations, net uptake of K^+, Cl^- and $H_2PO_4^-$ seems to be regulated by controlling the influx step. This is not to suggest that the rates of efflux of these ions are insignificant. For example, as K^+ influx declines, with increasing tissue concentration, influx may approach values close to efflux ($\sim$ 0.5 to 1.0 μmol g^{-1}h^{-1}) under steady-state conditions, so that net uptake (influx - efflux) may be quite low compared to rates in low-salt plants.

An important question regarding the regulation of ion uptake is the nature of the mechanism responsible for "down-regulating" influx as tissue concentrations rise. It might be argued that as ion concentrations rise, the free energy gradient against which the ion must be transported begins to limit influx. This would constitute *thermodynamic* regulation. Cram has made extensive investigations of the control of Cl^- influx and concluded that the decreased influx associated with increased tissue Cl^- (in carrot) could not be accounted for by the increased *thermodynamic* gradient ($\Delta\bar{\mu}_{Cl^-}$) against which the ion had to be transported. Similar conclusions were reached by Glass and Dunlop (1979) for K^+ influx in barley roots (see Table 7.3). Thus, the

Table 7.3.   Changes of $\Delta\bar{\mu}_{K^+}$ following treatment of low-salt barley roots with KCl solution to increase internal $[K^+]$. Although $[K^+]_{root}$ increased from 21 to 58 μmol K^+ g^{-1} fresh weight over the 4 h exposure, $\Delta\bar{\mu}_{K^+}$ across the plasmalemma actually increased because the electrical potential difference (E) became more negative. $\Delta\bar{\mu}_{K^+}$ was calculated as outlined in Section 5.3. (From Glass and Dunlop, *Planta*, 145: 395-397, 1979)

Period of treatment (h)	K^+ influx ($\mu mol\ g^{-1}h^{-1}$)	$[K^+]_{root}$ ($\mu mol\ g^{-1}$)	E (mV)	$\Delta\bar{\mu}_{K^+}$ ($kJ\ mol^{-1}$)
0	3.05	21	− 94	5.84
1	2.72	32	−118	4.59
2	2.16	48	−127	4.71
4	1.61	58	−138	4.11

influx of these ions is probably regulated via kinetic (metabolic) rather than by thermodynamic effects.

Many workers have reported that SO_4^{2-} influx is also regulated as tissue SO_4^{2-} or organic S levels of the root rise. Recently, however, Cram (1983) has suggested that SO_4^{2-} absorption in carrot discs is principally controlled at the tonoplast, and that plasmalemma influx is not subject to regulation in response to S levels of the tissue. Rather, plasmalemma SO_4^{2-} influx appears to rise, as does plasmalemma efflux, with increasing external SO_4^{2-} concentration. As a result, cytoplasmic SO_4^{2-} concentration results from the balance between influx and efflux, transport to the vacuole or xylem and reduction to organic S.

Numerous investigations by Jackson and his associates at Raleigh, North Carolina, have demonstrated that NO_3^- efflux may be extremely important in determining tissue NO_3^- levels. In the case of pearl millet, e.g. NO_3^- efflux increased to such an extent during darkness that it exceeded influx and there was net loss of N from the roots. My own studies with $^{13}NO_3^-$ indicate that influx is insensitive to NO_3^- content in barley roots even though net uptake is negatively correlated with root NO_3^- levels (see Table 7.4). This result points to NO_3^- efflux as a controlling factor. However Lee and his coworkers (1986) in England have found that $^{13}NO_3^-$ influx is increased if plants are deprived of NO_3^- for 3 days. Naturally, in the cases of NO_3^- and SO_4^{2-}, the reductions of these ions to organic N and S, respectively (Chapter 8), represent added complications for the regulation of tissue composition. The amino acid cysteine, for example, feeds back

Table 7.4. Patterns of net NO_3^- uptake and $^{13}NO_3^-$ influx in barley plants grown at various levels of NO_3^- in the external medium prior to measuring net NO_3^- uptake or $^{13}NO_3^-$ influx. (From Glass et al., *Plant Physiology*, 77: 379-381, 1985)

Level of NO_3^- pretreatment prior to experiments (mM)	Net Uptake $\mu mol\ NO_3^-\ g^{-1}h^{-1}$	$^{13}NO_3^-$ influx $\mu mol\ g^{-1}h^{-1}$
0.01	8.1	8.95
0.1	5.0	7.88
0.2	2.0	7.19
0.5	1.7	9.98

to the first enzyme of the pathway for SO_4^{2-} reduction, ATP sulfurylase (Fig. 8.19). There is clearly much more to be learned about the regulation of these fluxes.

REGULATION BY TRANSCRIPTIONAL CONTROL?

In microbial systems it has been established, on the basis of genetic, as well as physiological information, that two or more carrier systems may exist under transcriptional control. For example in the fungus *Neurospora* there are both high affinity and low affinity systems for Pi uptake. The high affinity system ($K_m = 2.43\ \mu M$) is produced at high levels in P-starved cells, but is much reduced in P-rich cells. Mutants lacking the high affinity system, as well as mutants which produce the latter constitutively (i.e., regardless of the availability of Pi), have been isolated. As well as this transcriptional control of influx there is also evidence to suggest a direct feedback inhibition of Pi transport by internal Pi. Regulation of the activity of these transport systems is therefore rather like the regulation of biochemical pathways, by processes of enzyme repression and direct (allosteric) feedback inhibition.

In higher plants such clear-cut mutants have not yet been identified, although mutants with defective NO_3^- transport systems have been obtained in *Arabidopsis*, a small annual in the cabbage family (Doddema & Feenstra, 1978). In *Lemna*, and in barley, K^+ influx has been demonstrated to respond quite rapidly to changes of internal $[K^+]$. The time scale of these events has been interpreted to suggest the operation of direct (allosteric) control mechanisms. I suggested (Glass 1976), on the basis of the relationships between K^+ influx and root $[K^+]$, that the K^+ transporter possessed four internal binding sites for K^+. Binding of K^+ to these sites was considered to reduce the affinity of the external binding site, thus diminishing influx. Release of these bound ions was suggested to relieve this allosteric inhibition of the K^+ transporter, leading to increased influx (Fig. 7.6). With the advantage of hindsight (12 years later), this model may be viewed as overly simplistic. However it is to be expected that models will be

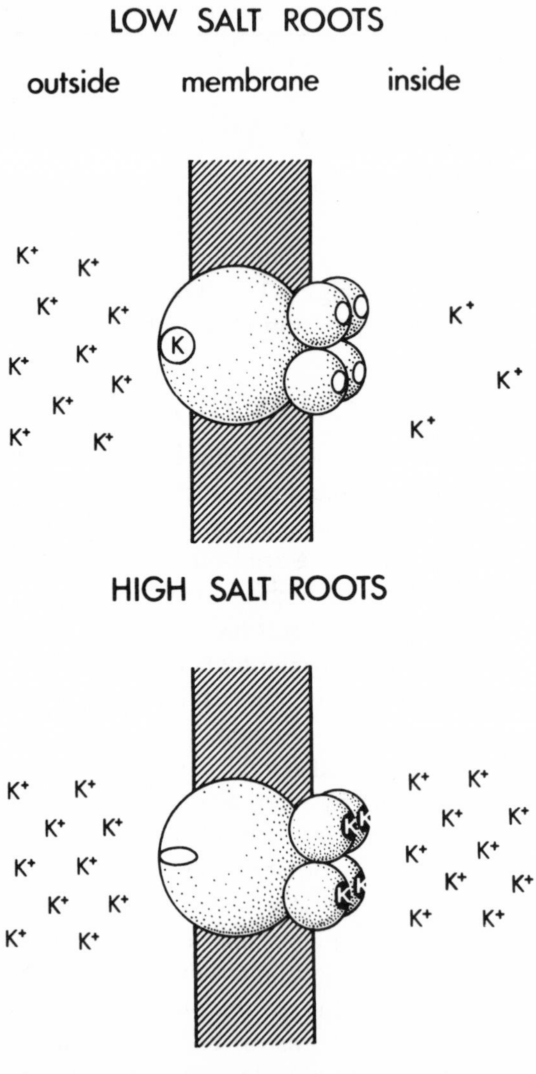

LOW SALT ROOTS

outside   membrane   inside

HIGH SALT ROOTS

Figure 7.6. A model for the allosteric regulation of potassium influx by internal K^+. In low-salt roots the 4 internal binding sites are free of bound K^+ and the external binding site has a high affinity for K^+. In high-salt roots the binding of K^+ to the allosteric sites causes a reduced affinity for external K^+. (From Glass, *Plant Physiology*, 58: 33-37, 1976)

discarded as we learn more about the system.

In most studies, e.g. Lee (1982), the withholding of particular nutrients (SO_4^{2-}, $H_2PO_4^-$, or Cl^-) has appeared to produce much greater effects upon the V_{max} for uptake than upon K_m values. This type of observation is most simply interpreted as the result of effects upon rates of carrier synthesis. However, as with the case for allosteric inhibition, the hard evidence for such mechanisms is meagre.

7.1d   CELLULAR PRIORITIES: CYTOPLASM BEFORE VACULE

The cytoplasm of plant cells is the site of intense biochemical activity. Bearing in mind the importance of the essential elements in cell metabolism (Chapter 8), several authors have stressed that the maintenance of a constant cytoplasmic milieu (vis à vis inorganic composition) must represent a high cellular priority. Thus it has been suggested that under conditions of nutrient deprivation the cytoplasmic composition can be maintained by adjusting transport rates across the plasma membrane and by reallocating nutrients which have been stored within the vacuole. For example, Lee and Ratcliffe (1983) have recently demonstrated that in pea roots cytoplasmic [Pi] remained constant at ~ 18mM when external Pi was varied. As total [P] of the tissue declined, cytoplasmic Pi was maintained at the expense of the vacuole whose [Pi] declined from 14 to 3 mM. Memon, (1985) has obtained similar findings for the allocation of K^+ between cytoplasm and vacuole (see Table 7.5). Conversely, when nutrients are in abundant supply considerable quantities of the latter are stored within vacuoles. This raises the question of the origin of the feedback signals which exert effects (both direct and indirect) upon the various transport systems. The greater size of the vacuole (approximately 90% of cellular volume) means that it represents a much larger reservoir of

Table 7.5.   Cytoplasmic and vacuolar concentrations (mM) of K^+ in barley roots (variety Compana) grown in hydroponic facilities maintained at 0.01 or 0.1 mM K^+. Also shown are the ratios for cytoplasmic: vacuolar [K^+]. (From Memon and Glass, *Journal of Experimental Botany*, 36: 1860-1876, 1985)

[K^+] of growth media (mM)	Cytoplasmic [K^+]	Vacuolar [K^+]	Ratio of concentrations
0.01	133	21	6.3
0.1	140	61	2.3

ions than the cytoplasm. The small volume of the cytoplasm and its high metabolic rate must make it more sensitive to perturbation by changes of external supply. Thus it would not be unreasonable for signals originating from cytoplasmic concentration to act upon both plasmalemma and tonoplast transporters, and by so doing maintain cytoplasmic homoeostasis. However, the vacuole is more than a repository for excess nutrients and harmful byproducts of metabolism. It plays an important role in the generation of turgor (Chapter 8). The constancy of average tissue concentrations of the elements, referred to previously, indicates that if the nutrients are available, vacuolar concentrations are regulated rather precisely too. It is difficult to see how this could be achieved without feedback signals from the vacuole to tonoplast and plasmalemma transporters. These ideas are summarized in Figure 7.7.

It cannot be overemphasized that the role of the root in obtaining inorganic nutrients is no more important than its role in transporting

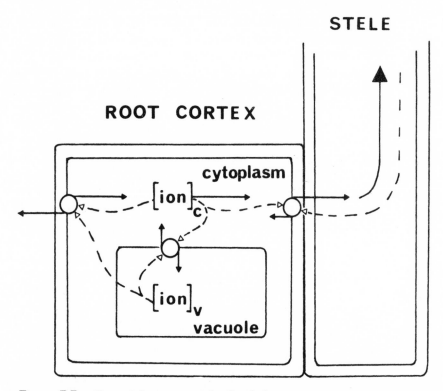

Figure 7.7.   Potential sources of feedback from various sources upon ion transporters in a diagrammatic representation of the root. Fluxes are shown by solid lines. Feedback is suggested by means of dotted lines. (From Glass and Siddiqi, *Advances in Plant Nutrition*, 1: 103-147, 1984)

these nutrients to the shoot. Under ideal conditions greater than 80% of absorbed nutrients may be translocated to the shoot. Hence it would be unreasonable to imagine that nutrient absorption by the root would not be responsive to the demands of the shoot.

Clearly this is an extremely complex problem and our present understanding of the details involved is quite limited. Since delivery of ions to the shoot involves absorption from soil solution and possibly also secretion into xylem vessels (Section 6.1c) both these processes must be sensitive to "signals" originating from the shoot. The most likely candidates for such signals are plant growth regulators such as auxin (IAA) or abscisic acid (ABA) or cytokinins. Experiments in which such compounds have been applied exogenously to root systems or even to decapitated seedlings indicate that both the absorption and translocation of solutes can be profoundly affected. For example Glinka (1980) has demonstrated that $4\mu M$ ABA causes increased transport of NO_3^- and K^+ to the xylem in decapitated sunflower plants (Fig. 7.8). Prior to the application of ABA the

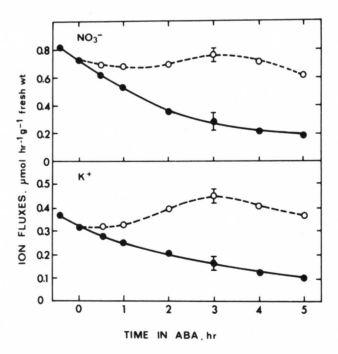

Figure 7.8. Effect of $4\mu M$ ABA on the fluxes of NO_3^- and K^+ to the xylem sap of decapitated sunflower plants. Two hours prior to the application of ABA plants were transferred to $CaSO_4$ solution. Thus, in controls (closed circles) there is a gradual decline of fluxes. Open circles represent ABA-treated plants. (Glinka, *Plant Physiology*, 65: 537–540, 1980)

plants were transferred in $CaSO_4$ solution so the increase of transport to the xylem (~300%) must have involved the removal of ions stored in the vacuoles of root cells. However, effects generated through exogenous application, though suggestive, may not represent the situation *in vivo*. The present state of our knowledge of these processes is well expressed in Figure 7.9 (from Pitman and Cram, 1977). Note the uncertainly expressed in the question mark on the line from the hormone box "to root".

7.2 Temperature

Soil temperature varies considerably on a seasonal basis and to a smaller extent even on a diurnal basis. Because of the water content of soil and its other physical properties, soil temperatures are, nevertheless, not subject to such wide fluctuations as those of the air. In harsh winters snow cover may insulate soil, thus maintaining its

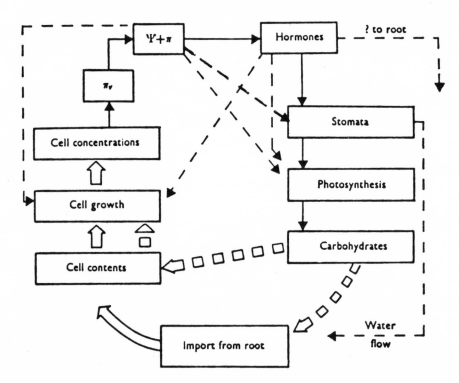

Figure 7.9. Relationships between the concentrations of solutes in leaf cells and other processes. Dotted lines demonstrate feedback processes. Ψ, water potential; π, osmotic pressure in the cell, π_v osmotic pressure in vacuole. (From Pitman and Cram, *Society for Experimental Biology Symposium*, 31: 391-424, 1977)

temperature well above that of the air. We have seen that nutrient uptake may be dependent on a source of metabolic energy, which is provided via respiration in roots. Since temperature exerts such a pronounced effect upon the rates of enzymatic reactions, ion transport should also be temperature sensitive. A quotient which is quite useful in discussing temperature effects, the Q_{10}, is defined as:

$$\frac{\text{Rate of a process at t}^{\circ}\text{C} + 10^{\circ}\text{C}}{\text{Rate of the process at t}^{\circ}\text{C}}$$

Clearly, a temperature sensitive process such as respiration would have a $Q_{10} = 2\text{-}3$, whereas solute diffusion, which is less sensitive to temperature, would have a Q_{10} close to unity. Barber (1972) showed that the Q_{10} for Pi uptake by beetroot tissue ranged from 3 at 0.1 mM Pi to below 1 above 20 mM. Plasmalemma fluxes of several other ions, including H^+, K^+, Cl^- and SO_4^{2-} in root tissues have Q_{10}'s between 2 and 3 at concentrations < 1 mM. Only Ca^{2+} influx, a passive process, has a Q_{10} close to 1. Considering that seedling germination and early growth commonly occur at temperatures in the 5 to 10°C range in springtime, the importance of temperature limitation of nutrient uptake might be considerable. In this regard, the almost universal choice of uptake temperatures in the 25-30°C range by experimenters using seedling roots, though convenient, may have been somewhat unwise. In those studies which have investigated temperature effects upon ion transport, the exposures to lower temperatures (generally in the range from 5-30°C), during relatively short uptake periods, have generally followed several days growth at 20 to 30°C. Measurements made under such conditions are analogous to the concentration perturbations which are imposed in most investigations of concentration dependence. By contrast, when plants are maintained at lower temperatures for several days, rates of nutrient uptake gradually increase so that they may approximate the rates obtained in plants maintained at much higher temperatures. According to the extent of temperature acclimation (like the concentration acclimation discussed earlier) this process tends to make ion uptake independent of temperature. This phenomenon was first reported by Clarkson and his associates in England. It was shown that roots of plants which had been pre-cooled at ~ 10°C for several days developed increased capacity for ion uptake, translocation of ions to the stele and increased flow of sap within the xylem. Chapin, working with various Alaskan plants, which experience mean soil temperatures ranging from 2 to 18°C in July, was able to demonstrate a clear-cut temperature compensation in Pi uptake (Table 7.6). A similar situation has been reported for K^+, Na^+ and Cl^- fluxes in *Chara*.

Table 7.6. V_{max} values for phosphate absorption by roots of various species, after growing at 5 or 20°C prior to the experimental period. (Chapin, *Ecology*, 55: 1180-1198, 1974)

Species	V_{max}* at 5°C after growing at:	
	5°C	*20°C*
Eriophorum angustifolium	0.58	0.14
Dupontia fischeri	0.78	0.97
Carex aquatilis	0.48	0.19
Scirpus microcarpus	0.34	0.06
Eleocharis palustris	0.26	0.07

*V_{max} is measured in μmol Pi g^{-1} freshweight h^{-1}

The capacity for temperature acclimation, therefore, appears to be widespread.

The increased ion transport capacity of cold-acclimated roots may be the result of increased synthesis of transporters. It is also possible, however, that changes in the lipid composition of membranes may play a contributory role. Such changes have been the focus of considerable debate among scientist involved in studying freezing resistance in crop plants. Unfortunately, no clear concensus has yet emerged.

7.3   Ionic Interactions

The absorption of particular inorganic ions can be strongly influenced by the presence of other (interacting) ion species in the external medium. Such interactions are possible between cations, e.g. K^+ and NH_4^+, Sr^{2+} and Ca^{2+}, or K^+ and Ca^{2-}; between anions, e.g. Br^- and Cl^-, or NO_3^- and Cl^-, and between cations and anions, e.g. NH_4^+ and NO_3^-, or Ca^{2+} and Cl^-. These ionic interactions may result from effects at the uptake step or may depend upon subsequent events. The former are generally more readily understood and will be discussed first.

The association between an ion and the putative carriers responsible for membrane transport depends upon the recognition of particular ions by specific carriers. Where two ions possess the same net charge and similar hydrated diameters (e.g. K^+ and Rb^+, with hydrated diameters of 0.54 and 0.51 nm, respectively) it might be anticipated that the carrier would be unable to discriminate between these ions. This appears to be the case, and K^+ uptake is, indeed, reduced by the presence of Rb^+ in the external medium because Rb^+ effectively competes with K^+ for the carrier's binding site. Epstein and his collaborators established that several such (competing) ion pairs exist e.g. Ca^{2+} and Sr^{2+}, $SO_4{}^{2-}$ and $SeO_4{}^{2-}$, and K^+ and Rb^+. By plotting 1/uptake versus 1/concentration (equivalent to the Lineweaver-Burk plot for enzymes), Epstein and Hagen (1952) were able to investigate the kinetics of competition between such pairs of ions.

If the Michaelis Menten equation (Section 5.5) is written in reciprocal form:

$$\frac{1}{v} = \frac{K_m + [S]}{V_{max} \bullet [S]}$$

it can be rearranged in the form of a straight line equation:

$$\frac{1}{v} = \frac{K_m}{V_{max}} \frac{1}{[S]} + \frac{1}{V_{max}}$$

Hence a plot of $\frac{1}{v}$ against $\frac{1}{[S]}$ (known as a Lineweaver-Burk plot) gives a straight line with slope $= \frac{K_m}{V_{max}}$ and Y intercept, (value of Y for $X = 0$) $= \frac{1}{V_{max}}$. When $\frac{1}{v} = 0$, it can be shown that $\frac{1}{[S]} = \frac{-1}{K_m}$. Thus the X value corresponding to $Y = 0$ gives the reciprocal of K_m. Figure 7.10A shows the uptake of Cl^- by barley roots in the presence (triangles) and absence (circles) of 1.0 mM Br^-. The data is taken from the published word of Elzam and Epstein, 1965. Clearly, the presence of Br^- severely reduces Cl^- uptake. Figure 7.10B shows the same data in the form of a Lineweaver-Burk plot. Despite the inhibitory effect of Br^- at low Cl^- concentrations the V_{max} for Cl^- uptake (4.7 μmol g^{-1}h^{-1}) was unchanged by the presence of Br^-. This is diagnostic for competitive inhibitors. In effect it indicates that as the 'true' substrate (Cl^-) is increased to infinite concentration ($\frac{1}{[S]} = 0$) then the effect of the competitive inhibitor is swamped out. Note that the inhibitor increases the estimate of K_m.

It would be a mistake to imagine that all ions interact at the uptake step in this way. For example, Epstein and his co-workers established that in low-salt barley roots Na^+ was without effect on K^+ uptake (below 1 mM) even though the concentration of Na^+ was ten times that of K^+. In *Agropyron*, absorption of Cl^- is unaffected by a 500-fold excess of $SO_4{}^{2-}$! In these two examples the transport systems for K^+ and Cl^- (respectively), must be highly specific for these ions.

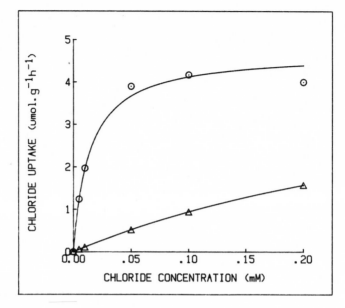

a.

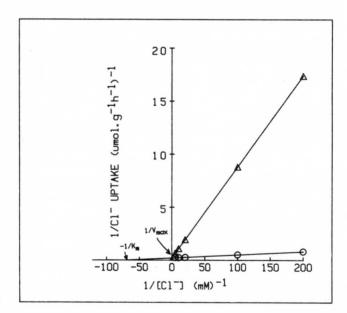

b.

Figure 7.10.   a. Chloride uptake by barley roots in the presence ($\triangle$) and absence of (o) of 1.0 mM Br^-.

b. A Lineweaver-Burk plot of the same data. Note that although Br^- alters the apparent K_m for Cl^- uptake, the V_{max} is unchanged. (From Elzam and Epstein, *Plant Physiol.* 40: 620-624, 1965)

7.3a COUNTER ION EFFECTS

Epstein and others (e.g. Hiatt) were also able to demonstrate that ion absorption in the low concentration range (< 1 mM) generally appeared to be independent of the nature of the counter ion. Thus Cl^- uptake is essentially the same from a solution of KCl or $CaCl_2$, despite the fact that K^+ is absorbed more rapidly than Ca^{2+}. Similarly K^+ absorption is unaffected by the substitution of SO_4^2 for Cl^- even though Cl^- is absorbed more rapidly than SO_4^{2-}. Unequal rates of cation and anion uptake would inevitably lead to charge separation, with significant consequences for membrane electrical potentials, were it not for counteractive (charge balancing) processes.

Over forty years ago Hoagland and Broyer observed that unequal rates of cation and anion absorption were accompanied by pH changes of the external medium. When cation uptake exceeds anion uptake the medium is acidified, while alkalinization of the medium accompanies excess anion uptake. These apparent ion exchanges (H^+ for excess cation and OH^- or HCO_3^- for excess anion) even formed the basis of a model for ion uptake proposed by Jacobson et al. in the 1950's. It was considered that the binding of cations (M^+) and anions (A^-) to the membrane carriers (a first step in transport) involved displacement (exchange) of H^+ or OH^-, respectively:

$$M^+ + HR \longrightarrow MR + H^+$$

and

$$A^- + R'OH \longrightarrow R'A + OH^-$$

where R and R′ represent cation and anion carriers, respectively.

It has also been clearly documented that excess K^+ absorption (and H^+ extrusion) leads to the accumulation of organic acid (typically malate) within roots. This malate is generated from the carboxylation of phosphoenolpyruvate (PEP) by the enzyme PEP carboxylase which is stimulated by alkaline pH. The product of this reaction is the compound oxaloacetate which is reduced by the enzyme malate dehydrogenase:

Conditions leading to excess anion uptake (e.g., Cl^- absorption from $CaCl_2$) lead to the breakdown of malate through the action of malic enzyme. Excess NO_3^- absorption, however, also leads to the formation of malate because NO_3^- reduction requires the consumption of H^+ equivalents (see Section 8.3). This situation (like excess cation uptake) should lead to alkalinization of the cytoplasm and hence, malate formation. All the ramifications of these responses have still not been entirely resolved. However, it has been proposed that H^+ extrusion (in exchange for K^+) causes cytoplasmic pH to increase. The enzyme PEP carboxylase is stimulated by alkaline pH with the result that a strong acid (malate) is generated which counteracts the tendency to displace cytoplasmic pH. At the same time malate provides anion equivalents for K^+ (so that internal charge balance is maintained).

Alkalinization of the external medium, associated with excess anion uptake, should lead to acidification of the cytoplasm, but this is retarded by the breakdown of malate. Despite these interesting explanations, recent studies using ^{31}P N.M.R. (nuclear magnetic resonance spectroscopy), a technique which makes it possible to measure cytoplasmic and vacuolar pH, failed to reveal changes of cytoplasmic pH during excess cation uptake. Nevertheless, these findings do not necessarily refute the above hypotheses. If the counter-active responses to a drift of cytoplasmic pH occur rapidly enough no *detectable* pH changes will result. This is not an unreasonable prediction considering the potent effects of pH on cellular processes.

Inorganic cation imbalance is not simply an artifact produced by allowing roots to absorb cations in excess of anions. Rather, even when cation uptake is balanced by uptake of nutrient anions (NO_3^- and SO_4^{2-}) the reduction of these ions (Section 8.3) generates an anion deficit which is associated with the synthesis of organic anions, particularly malate. The role of organic anions in balancing excess cation content of tissues is particularly evident in the cation/anion analyses (Table 7.7), reported by Kirkby (1968), for leaves of white

Table 7.7. Influence of the form of N available upon the cation-anion balance in white mustard leaves. Ion concentrations are expressed per unit of plant dry matter (DM). (Kirkby, *Soil Science*, 105: 133-141, 1968)

| | Cations (mEq 100 g^{-1}DM) | | | | | Anions (mEq 100 g^{-1}DM) | | | | | |
	Ca	Mg	K	Na	Total	NO_3^-	$H_2PO_4^{2-}$	SO_4^{2-}	Cl^-	Organic Acid	Total
NO_3^-	107	28	81	5	221	1	26	25	25	162	239
NH_4^+	72	22	40	7	141	1	25	25	31	54	136

mustard. Regardless of N source (NO_3^- or NH_4^+), the inorganic cation/anion deficit was made up almost entirely by organic acids.

We have observed that the independence of cation and anion uptake in the low concentration range may lead to compensatory ion fluxes and organic acid synthesis or degradation. In the high concentration range (> 1 mM) cation and anion absorption appear to interact to a much greater extent so that far less disparity in rates of cation and anion absorption are apparent. Epstein et al. (1963) showed that the rate of uptake of the more rapidly absorbed ion (e.g. Rb^+ in Rb_2SO_4 solution) tends to be reduced by the more slowly absorbed SO_4^{2-} ion (Fig. 7.11).

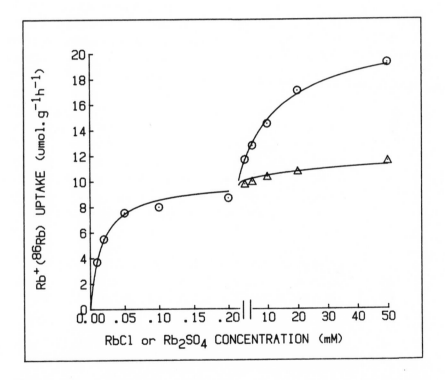

Figure 7.11. Rb^+ ($^{86}Rb^+$) absorption by excised barley roots as a function of external concentration of RbCl (o) or Rb_2SO_4 ($\triangle$). (From Epstein et al., *Proc. Nat. Acad. Sci.*, 49: 684-692, 1963)

7.3b IONIC ANTAGONISMS

Certain ion combinations e.g., NH_4^+ and K^+, and NO_3^- and Cl^- are well known for their negative interactions. The data of Table 7.3, for example, demonstrate the potent inhibitory effect of NH_4^+ on the accumulation of K^+ and Ca^{2-} in leaves of white mustard. In several other studies the inhibitory effect of NH_4 on K^+ uptake has been demonstrated at the initial uptake steps into the roots. Similarly, pronounced mutual antagonisms between Ca^{2+} and K^+ have been apparent since some of the early fertilizer trials reported at the beginning of the century.

Likewise, the accumulation of Cl^- has been shown to be strongly reduced by the presence of NO_3^- in the external medium (Fig. 7.5). The data of Hiatt and Leggett (Table 7.8) reveal that even quite moderate levels of NO_3^- cause substantial reductions of Cl^- accumulation, in both roots and leaves of barley. Recently, I have demonstrated that NO_3^- inhibits Cl^- influx at the plasmalemma in barley roots. However in roots which have received no NO_3^- this inhibitory effect is not apparent until about 3 h after exposure to NO_3^-. Interestingly, this corresponds quite well with the time required to activate ("induce") maximum NO_3^- uptake in plants which have previously received no NO_3^-.

It is difficult to imagine that ions as different as NH_4^+ and Ca^{2+}, or NO_3^- and Cl^-, could be competing for the same carriers. Recently some authors have proposed that it is not the carrier that is the target of competition, but rather $\Delta\mu_{H^+}$, the driving force for ion uptake via chemiosmotic transport mechanisms (Section 5.6). Clearly, depolarization of membrane potentials during ion uptake must reduce the driving force for ion uptake. In summary, the existence of ionic antagonisms are all too evident. The precise mechanisms of these interactions remain to be resolved.

Table 7.8. Accumulation of K^+, and Cl^- (μmol g^{-1}) by roots and shoots of barley plants as a function of the NO_3^- concentration of the growth medium. Cl^- was held constant at 1 mM. (Data of Hiatt and Leggett, 1971)

Growth Medium			Tissue Concentration			
$[NO_3^-]$	$[Cl^-]$	$[K^+]$	$[Cl^-]$		$[K^+]$	
			Shoot	Root	Shoot	Root
0	1	1.0	65	46	15	19
0.1	1	1.1	33	13	33	33
0.5	1	1.5	13	5	39	52
1.0	1	2.0	12	3	63	55
2.0	1	3.0	7	0	80	26
5.0	1	6.0	5	1	90	29

7.4 Light

The dependence of ion uptake on light in photosynthetic cells was first documented by Hoagland and coworkers in 1926. In subsequent years various algae and aquatic angiosperms (e.g. *Elodea* and *Vallisneria*) have been used extensively to investigate the nature of this dependence. There are several ways in which photosynthetically active radiation might influence ion fluxes. Light absorbed by the photosystems causes electron transport and coupled H^+ transport into the interior of the thylakoid cavities. These H^+ ion fluxes (Section 5.6) are balanced by counter fluxes of other cations (K^+ or Mg^{2+}) into the stroma. Secondly, the ATP generated during the light reactions may serve as the source of energy to bring about active ion fluxes. Finally, the carbohydrate generated via the dark reactions of photosynthesis may be subsequently oxidized, via respiratory reactions, to liberate the energy required for active transport.

As early as 1943, Arisz showed that the light-stimulation of Cl^- influx into *Vallisneria* leaves was independent of the presence of CO_2, thus eliminating possible stimulation of Cl^- uptake through increased carbohydrate supply. The similarity between the action spectrum for photosynthesis and Cl^- uptake by leaves of this same species led Lookeren-Campagne to propose a close linkage between photosynthesis and ion transport, possibly through increased availability of ATP.

More recent studies, particularly with algal cells, have revealed that in some cases ATP does not appear to represent the energy source for light-stimulated ion absorption. Moreover, in some cases a requirement for specific photosystems is indicated. For example, cation and anion fluxes in both *Nitella* and *Hydrodictyon* have been shown to depend upon different energy sources (ATP and photosystem II, respectively). Only a limited amount of work has been undertaken with leaves of terrestrial angiosperms, probably because of the technical difficulties associated with measuring ion fluxes in these tissues.

With the notable exceptions of the aquatic angiosperms, the acquisition of inorganic nutrients from the external environment by higher plants is almost exclusively confined to root tissues. These tissues are entirely dependent upon the translocation and subsequent oxidation of carbohydrate from photosynthetic organs for the energy required to absorb required nutrients.

Rates of absorption of inorganic nutrients by roots of intact plants commonly demonstrate diurnal patterns such as those illustrated in Figure 7.12 for NO_3^- uptake by ryegrass. These patterns of

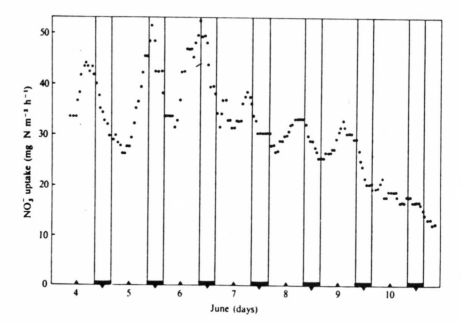

Figure 7.12. Hourly rates of NO_3^- uptake by perennial ryegrass plants growing in flowing nutrient solution containing 0.1 mg N l^{-1} during 7 days in June. Shaded segments represent the periods from sunset to sunrise. (From Clement, Hopper, Jones and Leafe, *Journal of Experimental Botany*, 29: 1173-1183, 1978)

nutrient absorption appear to depend upon a continuous supply of carbohydrate from the shoot. This conclusion is based upon the observed reduction of rates of nutrient absorption associated with treatments such as "girdling" of stem bark, or chilling of a segment of stem, to interrupt phloem translocation. Similarly, reductions of photosynthetic rates by shading aerial organs has frequently been shown to interfere with nutrient absorption by roots.

There can be no question that root activity is dependent upon the supply of carbohydrate from photosynthetic organs. However, it is not clear that the diurnal variations in absorption rates are the direct result of variations in carbohydrate supply to the root. Cram (1983) has emphasized that energy supplies for metabolic processes (such as ion transport) are typically generated at rates which match demand. Consequently, rather than ion uptake being limited by ATP supply, we might anticipate ATP supply being determined by rates of ion uptake. Naturally, active ion fluxes must decline when ATP synthesis or carbohydrate supply is blocked artificially (see Fig. 5.17). However, under natural conditions the increased export of photosynthetically-derived carbohydrate to the root is but one of several results of the increased shoot activity associated with the diurnal light exposure. Increased demand for inorganic nutrients which is correlated with this activity may be the cause of the enhanced rates of nutrient uptake. Thus, the carbohydrate supply may be essential as a source of energy for the latter process without necessarily being the regulating factor. Remember that most of the absorbed nutrients are translocated to the shoot. Hence, above all, shoot requirements should determine patterns of absorption. Pitman and Cram (1973) have emphasized the correlation between rates of K^+ transport to the shoots of barley plants and the relative growth rates of the latter (see Fig. 7.13). In this regard, it is interesting to note that in the study of NO_3^- uptake by ryegrass (Fig. 7.8) maximum rates of NO_3^- absorption lagged some 5 to 6 h behind maximum CO_2 fixation, and the lowest rates of NO_3^- uptake often occurred at mid-morning when CO_2 fixation rates were at their highest.

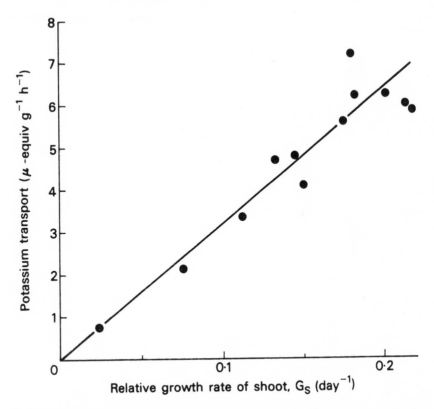

Figure 7.13. Relationships between the rates of K^+ transport to barley shoots and relative growth rates. Various relative growth rates were obtained by manipulating the photoperiod. (From Pitman, *Australian Journal of Biological Sciences*, 25: 905-919, 1972)

It appears that respiration rates also demonstrate a clear-cut diurnal periodicity. Respiration rates for roots of intact barley plants rise rapidly following the onset of illumination, reaching maximum values several hours into the light period (Fig. 7.14). Roughly 50% of the carbohydrate transported to the root in this study was used for respiration and 50% for growth. The author of this work (Farrar, 1981) concluded that the diurnal variation in root respiration was not controlled by rates of assimilate supply from the shoot. Some other (as yet unknown) factor originating from the shoot was suggested to be responsible for this effect.

It is clear that photosynthetic rates, shoot growth, carbohydrate export, nutrient demand, root respiration, root growth and nutrient absorption are all influenced in a co-ordinated manner in response to light. The integration of these processes and in particular the integration of root and shoot activity is complex and poorly understood. A role for growth regulators would seem obvious in this context but as yet we are far from an understanding of the details of how this might operate.

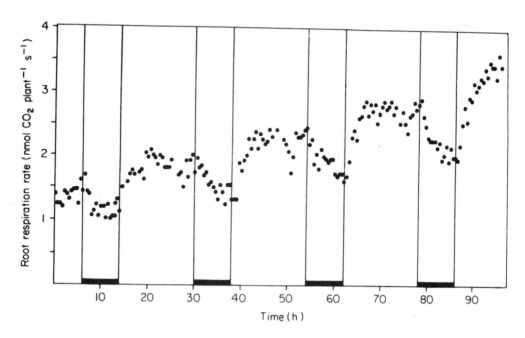

Figure 7.14.   Pattern of root respiration for barley plants between 14 and 18 days old. The solid bars represent 8 h dark periods. (From Farrar, *Annals of Botany*, 48: 53-63, 1981)

7.5 Water

A continuous supply of water is absolutely essential for normal plant growth and development. The role of water is crucial for so many physiological processes and the interdependences of these processes is so complex that is it sometimes difficult to decide which are the primary effects of water deficits and which are of secondary or even tertiary origin. There are, for example, profound changes in both the physical and biological properties of soil associated with reduced water content. Excessive soil water should not be overlooked either, for the flooding of soil may lead to leaching of nutrients and anaerobic soil conditions. In addition, water deficits exert pronounced effects upon the plant itself. Cell growth, cell wall synthesis, protein synthesis, carbon assimilation, respiration and many other physiological processes have been shown to be inhibited as tissue water potential is reduced.

EFFECTS UPON SOIL PROPERTIES

As soil dries after heavy rains the first water is lost by gravitational drainage. When this water has drained soil is said to be at field capacity (F.C.). From F.C. (-0.1 to -0.3 bar) to permanent wilting percentage (P.W.P.), at about -15 bars, this soil water is considered to be available for plant absorption. However, bulk flows of water toward the root are dramatically reduced as the soil water potential decreases toward P.W.P.. This is reflected in the substantial decrease of hydraulic conductivity (L_p) associated with reduced water content. L_p has been reported to decline from 10^{-5} cm s^{-1} at 50% soil water content (by volume) to below 10^{-11} cm s^{-1} at 10% soil water. This reduced conductivity is associated with the emptying of most of the soil water from the larger capillaries. The flow of water in soil capillaries is given by Poiseuille's Law:

$$J = \frac{\pi r^4 \Delta P}{8\eta l}$$

where r is the radius of the capillary, ΔP is the pressure gradient (due to a difference of water potential), η is the viscosity of the solution and l is the length of the channel. The flow of soil solution toward the root experiences little resistance as it proceeds via the larger capillaries. However, when this water is gone the remaining water (that which is adsorbed to the soil surfaces lining the larger pores and that filling the cavities of smaller capillaries) is much less available. Poiseuille's law dictates that a 50% decrease of capillary radius would reduce J to 1/16 of its original rate. A concomitant of this reduced rate of water flow to the plant is that the provision of inorganic nutrients, by bulk flow (Section 3.5), is diminished to an equivalent extent as soil progressively dries.

In addition, the path length for diffusion of ions increases as the water film adhering to soil particles decreases in thickness. Furthermore, the shrinking of this layer brings ions into closer proximity with the charged surfaces of soil particles. The combined effect of these factors is to diminish the rate of supply of nutrients, such as K^+ or Pi, which arrive at the root surface largely by diffusion.

Water deficits also exert a pronounced effect upon the growth and activity of soil micro-organisms. The latter are critical for the recycling of inorganic nutrients through their action upon organic matter and soil minerals. The importance of soil micro-organisms in relation to nitrogen cycling and phosphate absorption is emphasized in Sections 3.1b and 4.5a, respectively.

EFFECTS UPON ROOT ACTIVITY

Both deficits and excesses of soil water have a detrimental influence upon root growth and nutrient absorption. Successful root growth of numerous species in hydroponic facilities attests to the fact that water, *per se*, has little deleterious effect upon root growth. Rather it is the anaerobic conditions which prevail in flooded soils which are responsible for inhibiting root growth. The absorption of K, N, Mg, Ca and P by corn plants growing in non-aerated (water-logged) soil was found to be 30%, 70%, 80% and 90% and 130%, respectively, of rates of uptake in control (aerated) soil.

As soil water content falls below F.C., root growth declines well before P.W.P. is reached. As a consequence, together with the previously discussed effects of water deficits, the plant may become severely deprived of inorganic nutrients. The reduction of transpiration, photosynthesis and growth, associated with declining soil moisture levels, all result in a reduction of plant "demand" for inorganic nutrients. However, as stated earlier, it is difficult to separate these sorts of effects from the more direct effects of water deficits on ion transport. Several laboratory studies have demonstrated that the uptake of Pi, K^+, and other ions is inhibited when the water potential of the nutrient solution is reduced by means of non-penetrating organic solutes such as mannitol. This is not an ideal simulation of water stress in soils because treatments such as mannitol are associated with decreased osmotic potentials rather than the reduced matric potentials (due to adsorption of water on soil colloids) which accompany drying out of soils. Nevertheless, in thermodynamic terms, i.e. on the basis of the reduction of free energy of water, matric potentials and osmotic potentials are indistinguishable. The results of such experiments may therefore still serve as useful indications of the situation in soils. The measurement of total ^{86}Rb uptake from labeled soils of varying water content by corn seedlings has established extremely strong correlations between uptake and water content (Fig. 7.15). However, in experiments of this sort the effects upon soil properties were not separated from direct effects upon ion transport.

An analysis of the influence of soil water content upon rates of

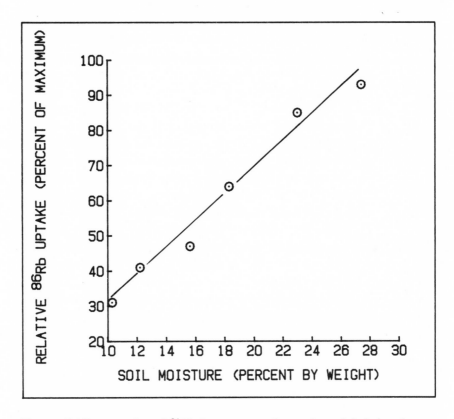

Figure 7.15. Uptake of ^{86}Rb by corn seedlings, from labeled soil, as a function of soil moisture. Correlation coefficient for the regression line was 0.99. (From Danielson and Russel, *Soil Science Society of America Proceedings*, 21:3-6, 1957)

nutrient movement up to the root surface (in onion) and upon the roots' absorbing power revealed that both processes were reduced in dry soils. However the relative reductions varied according to the nutrient concerned. In the cases of K^+ and Cl^- there were comparatively larger effects upon transport up to the root surface. By contrast, for Pi there were more pronounced effects upon absorbing power.

As described in the Introduction to this chapter, physiologists typically investigate the effects of environmental factors one at a time. In growth cabinets, where these factors are readily controlled, this approach is quite feasible. In nature, however, environmental variables do not change independently of each other, and in addition there are intrinsic developmental variations during the season.

For example, in early spring when seedlings germinate in the field, temperatures and light intensities tend to be low, photoperiods are short while water and inorganic nutrients tend to be abundant. Rates of ion absorption by plants at this stage are extremely high (on a per g. root basis). However, root biomass increases exponentially during the early phases of growth so that total uptake (on a per plant basis) increases to its highest level later on into the season.

Nevertheless, as described in Chapter 6, high rates of absorption during the early vegetative growth period lead to the accumulation of nutrient reserves within tissues. As development proceeds into the reproductive period, ion absorption by roots still continues but at reduced rates. In the field declining rates of ion absorption may result from the depletion of soil reserves, drying of upper soil layers as well as intrinsic developmental influences. During the reproductive phase mobile nutrients are translocated to the developing fruits and ion concentrations in vegetative organs tend to decline. This is not the case for immobile elements such as Ca^{2+} which are not subject to retranslocation.

This pattern of nutrient absorption and translocation has been clearly revealed in studies of K^+ and Pi influx by winter wheat during its life cycle (Fig. 7.16). In this study by Erdei et al. (1983) both soil and hydroponically grown plants demonstrated similar seasonal patterns. However, K^+ influx was more strongly reduced during winter months. In spring there was an initial increase of influx prior to a gradual decline (Fig. 7.16b).

Summary

Short-term studies have identified numerous environmental factors which are capable of influencing the rates of ion absorption by plant roots. Among these, external concentration, temperature, ionic inter-

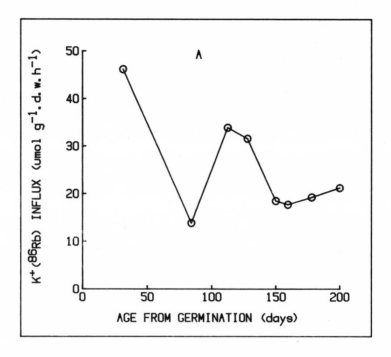

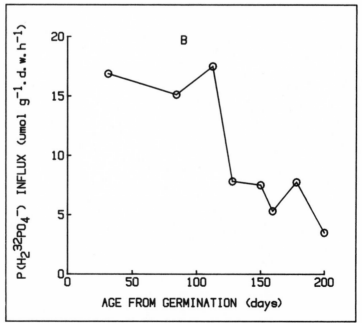

Figure 7.16.   Variation of Pi(a) or K^+(b) influx into roots of winter wheat during the life cycle. Influx is measured as μmol K^+ (^{86}Rb) or Pi ($H_2{}^{32}PO_4{}^-$)g^{-1} dry weight (D.W.)h^{-1}. (From Erdei et al., *Physiologia Plantarum*, 58:131-135, 1983).

actions, light and water have been extensively studied. Some of these, for example, concentration and temperature, may exert direct effects upon rates of ion absorption. In addition, there may be indirect effects due to alterations in growth rates or even effects upon soil properties.

Nevertheless, long-term studies have demonstrated that plants possess a remarkable capacity to acclimate to certain of the above environmental variables. As a result they tend to be relatively independent of environmental perturbations. However, when the latter are extreme, plants may be unable to accommodate and the absorption of nutrients and growth may be impaired. At the molecular level, various feedback mechanisms have been proposed to account for the observed responses to environmental variations.

Further Reading

Cram, W.J. 1976. Negative feedback regulation of transport in cells. The maintenance of turgor, volume and nutrient supply. In *Encyclopedia of Plant Physiology, New Series, Vol. IIA*, eds. U. Lüttge and M.G. Pitman. Berlin: Springer-Verlag.

Clarkson, D.T., and Deane-Drummond, C.E. 1980. Thermal adaptation of nitrate transport and assimilation in roots? In *Nitrogen as an Ecological Factor*. Proceedings of the British Ecological Society Symposium. eds. I. Rorison and J.A. Lee, pp. 1-20.

Deane-Drummond, C.E. 1982. Mechanisms for nitrate uptake into barley (*Hordeum Vulgare* c.v. Fergus) seedlings grown at controlled nitrate concentrations in the nutrient medium. *Plant Science Letters*, 24:79-89.

Fitter, A.H., and Hay, R.K.M. 1981. *Environmental Physiology of Plants*. London: Academic Press.

Glass, A.D.M. 1983. Regulation of ion transport. *Annual Review of Plant Physiology*, 34:311-26.

Glass, A.D.M. and M.Y. Siddiqi. 1984. The control of nutrient uptake rates in relation to the inorganic composition of plants. *Advances in Plant Nutrition*, 1:103-147.

Hiatt, A.J., and J.E. Leggett. 1974. Ionic interactions and antagonisms in plants. In *The Plant Root and its Environment*, ed. E.W. Carson, pp. 101-134. Charlottesville; University Press of Virginia.

Lee, R.B. 1982. Selectivity and kinetics of ion uptake by barley plants following nutrient deficiency. *Annals of Botany*, 50: 429-49.

Pitman, M.G., and Cram, W.J. 1977. Regulation of ion content in whole plants. In *Integration of Activity in the Higher Plant*, ed. D.H. Jennings, pp. 391-424. Society for Experimental Biology Symposium 31. Cambridge: Cambridge University Press.

Raven, J.A. 1977. Regulation of solute transport at the cell level. In *Integration of Activity in the Higher Plant*, ed. D.H. Jennings, pp. 73-99. Society for Experimental Biology Symposium 31. Cambridge: Cambridge University Press.

8

Biological Roles of the Inorganic Elements

The Periodic Table of chemical elements contains ninety two naturally occurring chemical elements (Fig. 8.1). It is quite remarkable to consider that plant life is based upon a mere sixteen of these. It is even more interesting to ask why these elements were selected. They are certainly not the most abundant elements. For example, Si is much more available than C; Al is considerably more common than K. George Wald (1962) has stressed that the choice of particular elements for particular biological functions was governed by their unique chemical properties.

In this chapter we examine the biological roles of the elements which are constituents of plant cells, and the consequences of inadequate supplies of these elements for plant growth and development. In addition, we shall consider the unique chemical and physical properties which enable the elements to perform their particular biological functions.

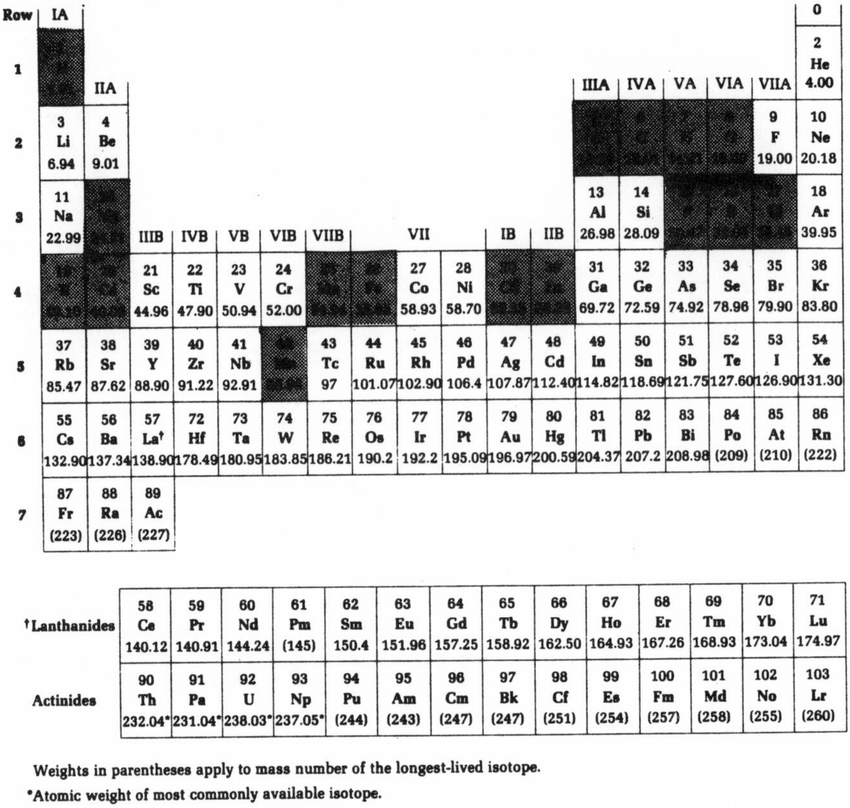

Row	IA																	0
1																		2 He 4.00
		IIA											IIIA	IVA	VA	VIA	VIIA	
2	3 Li 6.94	4 Be 9.01															9 F 19.00	10 Ne 20.18
3	11 Na 22.99		IIIB	IVB	VB	VIB	VIIB		VII		IB	IIB	13 Al 26.98	14 Si 28.09				18 Ar 39.95
4			21 Sc 44.96	22 Ti 47.90	23 V 50.94	24 Cr 52.00			27 Co 58.93	28 Ni 58.70			31 Ga 69.72	32 Ge 72.59	33 As 74.92	34 Se 78.96	35 Br 79.90	36 Kr 83.80
5	37 Rb 85.47	38 Sr 87.62	39 Y 88.90	40 Zr 91.22	41 Nb 92.91			43 Tc 97	44 Ru 101.07	45 Rh 102.90	46 Pd 106.4	47 Ag 107.87	48 Cd 112.40	49 In 114.82	50 Sn 118.69	51 Sb 121.75	52 Te 127.60	53 I 126.90	54 Xe 131.30
6	55 Cs 132.90	56 Ba 137.34	57 La† 138.90	72 Hf 178.49	73 Ta 180.95	74 W 183.85	75 Re 186.21	76 Os 190.2	77 Ir 192.2	78 Pt 195.09	79 Au 196.97	80 Hg 200.59	81 Tl 204.37	82 Pb 207.2	83 Bi 208.98	84 Po (209)	85 At (210)	86 Rn (222)
7	87 Fr (223)	88 Ra (226)	89 Ac (227)															

†Lanthanides	58 Ce 140.12	59 Pr 140.91	60 Nd 144.24	61 Pm (145)	62 Sm 150.4	63 Eu 151.96	64 Gd 157.25	65 Tb 158.92	66 Dy 162.50	67 Ho 164.93	68 Er 167.26	69 Tm 168.93	70 Yb 173.04	71 Lu 174.97
Actinides	90 Th 232.04*	91 Pa 231.04*	92 U 238.03*	93 Np 237.05*	94 Pu (244)	95 Am (243)	96 Cm (247)	97 Bk (247)	98 Cf (251)	99 Es (254)	100 Fm (257)	101 Md (258)	102 No (255)	103 Lr (260)

Weights in parentheses apply to mass number of the longest-lived isotope.

*Atomic weight of most commonly available isotope.

Figure 8.1. The Periodic Table of Elements. Shaded areas identify the essential elements.

8.1 Inorganic Constituents of Plant Cells

Since the time of de Saussure (1804) scientists have analyzed plant ashes to ascertain the nature and quantities of the chemical elements that are contained within the living plant. Although Liebig believed implicitly that elemental composition reflected the plant's exact quantitative requirements, which he thought should be used to dictate fertilizer practice, it is now recognized that many elements may be included within plant tissues even though they serve no essential function.

The analysis of plant tissues generally involves first drying the material for 24 h at 80°C, followed by grinding to a fine consistency. Subsequently, the resulting powder may be ashed at high temperature

(500°C) to convert organic constituents to their respective oxides. The ash which remains consists solely of inorganic elements and is then usually digested in strong acids prior to determination of the elemental composition using an atomic absorption spectrometer. We dried six week-old barley (*Hordeum vulgare*) plants whose fresh weights averaged 16.82±4.29 g per plant. After drying, the weight was reduced to 2.2±.53 g per plant. When the dried material had been ashed its weight amounted to 0.24±0.03 g (11% of dry weight and 1.4% of fresh weight). Thus, 16.82 g of fresh plant material contained only 1.4% of inorganic matter. Such a small percent may appear almost trivial and it is perhaps not surprising that van Helmont considered the 2 oz weight loss of the soil in his famous experiment to be due to experimental error! Nevertheless this small inorganic component is vital. Without its input plant growth is severely retarded (see Fig. 8.2).

Soil scientists and agriculturalists often present elemental concentrations of plant tissues in parts per million (1 p.p.m. = $1\mu g\ g^{-1}$) of plant dry weight. By contrast plant physiologists, who are usually more concerned with 'physiological' concentrations, often express tissue concentrations in μequivalents or μmoles per g fresh weight (g^{-1}.f.w.) or even as molarities, on the assumption that 1 g.f.w. of tissue is roughly equivalent to a volume of 0.8 cm^3 of water.

Figure 8.2. Corn plants after several weeks growth in solutions containing all the essential elements (complete) or in media lacking N, P, or K, respectively. (Glass, unpublished)

8.1a ESSENTIAL AND NONESSENTIAL CONSTITUENTS

For most plants approximately sixteen elements are considered to be essential. Arnon and Stout (1939) have provided a valuable definition of an essential element as one which participates directly as an indispensable requirement for the normal life cycle. The group of sixteen consists of C, H, O, N, P, S, K, Ca, Mg, Fe, Mn, Cu, Zn, Mo, B and Cl. Eight of the essential elements are absorbed in the form of oxides (e.g. CO_2 and SO_4^{2-}) while seven are taken up as metallic ions such as K^+ and Mg^{2+}. Add to this Cl^- (a halide) and NH_4^+ (a hydride) and the list is complete. Table 8.1 lists the forms in which the elements are available to plants. These elements are required by all

Table 8.1.   Chemical forms of the essential elements which are available to plants.

Element	Available form	
	Oxides	
C	CO_2	- carbon dioxide
H	H_2O	- water
O	CO_2, H_2O	
N	NO_3^-	- nitrate
S	SO_4^{2-}	- sulfate
P	$H_2PO_4^-$, HPO_4^{2-}	- phosphate
Mo	MoO_4^{2-}	- molybdate
B	BO_3^-, $B_4O_7^{2-}$	- borate
	Metallic ions	
K	K^+	- potassium
Ca	Ca^{2+}	- calcium
Mg	Mg^{2+}	- magnesium
Fe	Fe^{2+}	- ferrous (State II) iron
Mn	Mn^{2+}	- manganese
Cu	Cu^{2+}	- copper
Zn	Zn^{2+}	- zinc
	Halide	
Cl	Cl^-	- chloride
	Hydride	
N	NH_{4+}	- ammonium

plants, but in addition, other elements may be required by specific plant groups. For example, legumes, such as beans and peas, require Co when they depend upon nitrogen-fixing root nodules for their source of N; grasses and horsetails (*Equisetum* species) need Si; plants such as *Atriplex vesicaria* (Fig. 8.18), which grow in saline habitats require Na. Other elements may be contained within plant tissues without (apparently) serving any particular function. For example, some soils contain high concentrations of potentially toxic elements (Se, Hg, Cd) either from natural sources or as a result of industrial pollution. Analysis of the tissues of plants growing in such soils reveals that while some species or varieties can exclude these toxic elements, others accumulate and tolerate high internal concentrations, apparently without detrimental effects. It is evident that the latter plants have evolved mechanisms to detoxify these chemicals (see Chapter 9 for further details).

The histogram in Figure 8.3 provides a quantitative representation of the elemental composition of a corn plant. This is intended

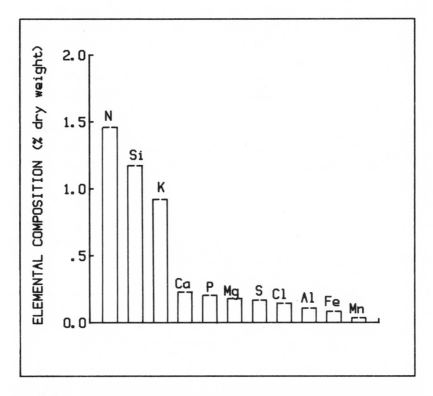

Figure 8.3.   Elemental composition of corn plants expressed as percent of dry weight. (From Miller, in *Plant Physiology*, 3rd ed. McGraw-Hill, 1938)

only as a guide to the elemental composition of higher plants, for percent composition may vary according to the plant species, and growing conditions. Newton e.g. grew sunflowers, beans, wheat and barley under greenhouse conditions in soils of identifical composition. Table 8.2 shows that the elemental compositions of these plants were quite different. Collander (1937, 1941) grew 21 species of higher plants in identical solutions, and observed that although there was hardly any variation in K^+ concentration among these species, the Na content ranged from 0.5 to 30% of total cation content. Even within the same species, e.g. white mustard (Table 7.7) the elemental composition of plants changes substantially when the N supply is switched from NO_3^- to NH_4^+.

8.2   Functions of the Elements

8.2a.   OSMOTIC PHENOMENA AND OSMOREGULATION

WALL-LESS CELLS

Many single celled algae possess no cell wall whatsoever. *Dunaliella* and *Ochromonas* (Fig. 8.4) are two examples of such algae, both of which are found in marine habitats. Protoplasts (Fig. 8.4), which physiologists can now generate from leaves or roots by digesting away the cell walls with appropriate enzymes, are also wall-less. When these wall-less cells are transferred to distilled water they swell rapidly and burst (lyse). When transferred to very concentrated solutions they shrink. These rapid volume changes are the result of flows of water along its free energy gradient (see Section 3.5). This phenomenon, known as osmosis, can be demonstrated equally well in

Table 8.2.   Percent of Ca, K, Mg, N and P (as % dry weight) in the tops of sunflowers, beans, wheat and barley. Plants were grown in greenhouses in identical soil. (From Newton, *Soil Science*, 26: 85-91, 1928)

Plant	Percent dry weight in tops				
	Ca	*K*	*Mg*	*N*	*P*
Sunflower	1.68	3.47	0.73	1.74	0.08
Bean	1.46	1.19	0.57	1.48	0.05
Wheat	0.46	4.16	0.23	2.26	0.06
Barley	0.68	4.04	0.29	1.94	0.13

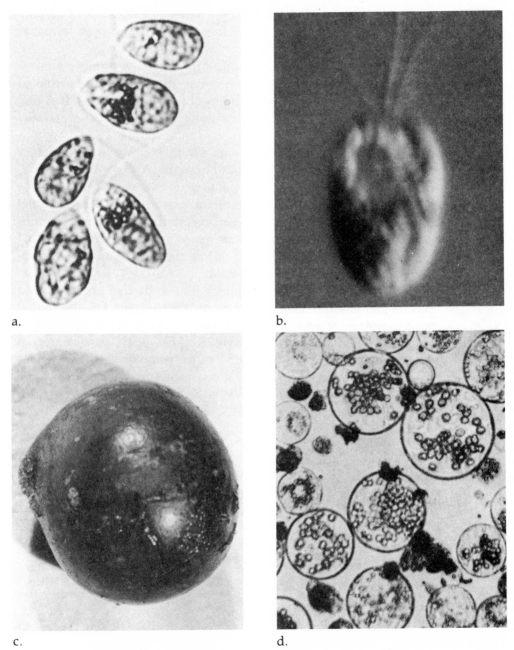

a.

b.

c.

d.

Figure 8.4.   Examples of wall-less cells:
(a) *Dunaliella* (Photograph courtesy of Dr. Emilia Klut); (b) *Platymonas* (Photograph courtesy of Dr. J. Stein); (c) Giant-algal cell (Valonia) which has been widely used for studies of turgor regulation. (Photograph courtesy of Dr. J. Stein) (d) protoplasts from bean leaves (Isolated by means of enzymic digestion of the cell walls)

non-living systems. Indeed, the laws governing these osmotic phenomena were worked out (by Pfeffer, 1899) using copper ferrocyanide membranes embedded in porcelain pots (Chapter 2).

All that is necessary to demonstrate osmosis is to place a membrane which is permeable to water, but relatively impermeable to solute, between solutions of different solute concentrations. It is well known that water will then flow toward the more concentrated solution. At equilibrium, when solute concentration is equal on both sides of the membrane, there is no net flux of water. In order to prevent swelling or shrinking of wall-less cells, whether they be algae, protoplasts or red blood cells, it is therefore necessary that they be held in solutions whose solute concentrations are equivalent (isotonic) to those of the cell. Therefore, when experimenters isolate leaf protoplasts they commonly use extraction solutions containing 0.6 M mannitol. Inadvertent dilution of this solution leads to lysis of protoplasts and a drastic reduction of yield.

The osmotically active solutes within cells vary according to species and environment, but typically they include organic solutes such as sugars, organic acids, amino acids and other nitrogen derivatives and inorganic ions, notably K^+, Cl^- and (particularly in marine organisms) Na^+. Table 8.3 lists the osmotically active solutes in leaves of bean (*Phaseolus*) grown with and without (+ and −) an adequate potassium supply. Note how the other solutes compensated for the lack of sufficient K when plants were grown without adequate potassium.

Table 8.3. Osmotically active solute concentrations (mM) in *Phaseolus* leaves grown with (+) or without (−) an adequate K supply. (From Mengel and Arneke, *Physiologia Plantarum*, 54: 402-408, 1982)

Solute	Older leaves +	Older leaves −	Younger leaves +	Younger leaves −
Organic anions	94	129	106	116
Organic N	335	290	408	433
K^+	101	25	135	50
Mg^{2+}	12	21	12	24
Ca^{2+}	51	66	21	47
Cl^-	32	33	21	27
TOTAL	625	564	703	697

It was stressed in Section 3.5 that spontaneous flows of matter always occur along gradients of free energy. It follows, then, that when water flows toward the more concentrated solution in osmotic systems, the free energy of water in the concentrated solution must be lower than that in the dilute solution. Why should this be so?

Although water (H_2O) is a covalently bonded molecule there is polarization of the electron pairs between oxygen and hydrogen (Fig. 8.5). Electron attracting (electronegative) O becomes slightly negatively charged, while H atoms bear a slight positive charge. The

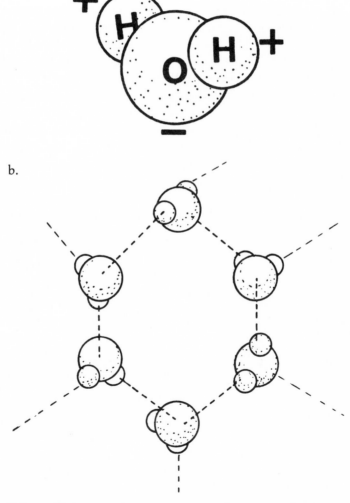

Figure 8.5a. Diagrammatic representation of a water molecule.
    b. One type of lattice structure of water due to hydrogen bonding.

attraction between a positively charged H of one molecule and a negatively charged O of a nearby water molecule forms a weak but important electrostatic bond, termed a hydrogen bond. Even at room temperature water is not composed of free water molecules but, rather, contains numerous lattices of hydrogen bonded H_2O molecules (Fig. 8.5b).

When a solute dissolves in water some of the polarized H_2O molecules become oriented around the solute. The layer of H_2O molecules which is closely bound to the solute is called the hydration shell. For example, non-hydrated Ca^{2+} has a diameter of 0.21 nm, whereas hydrated Ca^{2+} (Table 8.4) has a diameter of 0.56 nm. The lower free energy of a solution (compared to pure water) arises from the increase of entropy of the system. Consider what happens as a crystal of KCl dissolves in water. The solute molecules are greatly randomized leading to increased entropy and as a consequence a lowered free energy of the solution.

Although *Dunaliella* or other wall-less cells burst when subjected to excessive osmotic shock, they can survive gradually imposed osmotic changes by adjusting internal concentrations of solutes. Such gradual changes are not unusual in their natural habitats because of fluctuations in salinity. Consider the situation when a cell swells in a dilute (hypotonic) solution. The increase in volume is rapid because the hydraulic conductivity of the membrane is high. To restore volume to its original value the cell must remove water, either directly by pumping out water (as with contractile vacuoles) or indirectly by reducing internal solute concentrations. The latter method would cause an osmotic flow of water out of the cell and bring cell volume back to "normal".

Many fresh water algae use contractile vacuoles to secrete water from the cell. However, it is apparent that during volume regulation

Table 8.4.   Hydrated and non-hydrated diameters of some inorganic ions.

Ion	Hydrated diameter (nm)	Non-hydrated diameter (nm)
Cl^-	0.50	0.36
$Rb+$	0.51	0.30
K^+	0.53	0.27
NH_4+	0.54	0.29
Na^+	0.76	0.20
Ca^{2+}	0.56	0.21

wall-less cells make use of existing solute gradients to drive ions out of the cell along their thermodynamic gradients. This has been clearly documented for red blood cells and the evidence suggest that similar processes may also occur in *Dunaliella*. In nature there is generally a high gradient of K^+ concentration between the cell and the external medium. Thus, when the cell needs to "off-load" solute the permeability of the plasma membrane to K^+ is increased (Fig. 8.6). This results in a release of K^+, while Cl^- tends to follow passively (to maintain charge balance). The decrease of solute concentration within the cell is accompanied by an egress of water (by osmosis) and thus volume is restored.

When a cell shrinks as a result of being immersed in a more concentrated (hypertonic) solution it must increase its solute concentration, so that water will re-enter and restore volume. In *Dunaliella*, as in red blood cells, Na^+ permeability of the cell membrane increases and, because of the favourable gradient for Na^+ entry, the solute content of the cell is elevated (see Fig. 8.6), water follows, with concomitant restoration of volume.

Some text-books refer to the osmotic roles of the inorganic elements as being non-specific because, theoretically, the same concentration of any solute exerts the same osmotic effect. However, in practice, it is evident that specific solutes have been selected for osmoregulatory functions. One reason for this specificity is that high concentrations of certain elements interfere with metabolic processes and may even prove toxic to plant cells.

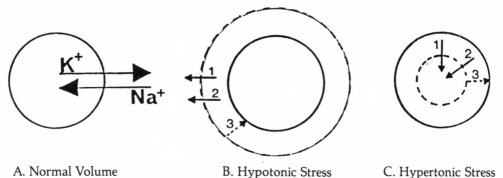

A. Normal Volume   B. Hypotonic Stress   C. Hypertonic Stress

Figure 8.6. Fluxes of Na^+ and K^+ associated with volume restoration in wall-less cells. Diagram A shows normal volume and existing gradients of Na^+ and K^+. In B. cells have swollen to a larger volume (dotted lines) in hypotonic media. Increased permeability to K^+ leads to solute loss (1) water loss (2) and restoration of volume (3). In C. cell shrinkage in hypertonic media leads to increased Na^+ permeability and associated solute absorption (1), water entry (2) and restoration of volume (3).

Notwithstanding the importance of inorganic salts in osmoregulation, organic solutes also play an important role. Inorganic ions are (energetically speaking) a "cheap" form of solute compared to organic molecules. However, although the vacuole can tolerate extremes of pH, high concentrations of Ca^{2+} salts, and high Na:K ratios, the cytoplasm appears to be much more sensitive. This is not surprising considering the many biochemical processes which occur within the latter compartment. Hence, some researchers consider that the use of inorganic ions to osmoregulate, though "cheap" and convenient, may nevertheless be only a "stop gap" so far as the cyptoplasm is concerned. In *Dunaliella* e.g., hypertonic stress leads to elevated cellular $[Na^+]$ as predicted in Fig. 8.6c. However within 2h $[Na^+]$ is restored to its original value (Fig. 8.7). In the meantime glycerol levels build up and take up the osmotic "slack". Thus, when *Dunaliella viridis*, previously equilibrated in 1.5 M NaCl, was transferred to 4.0 M NaCl, glycerol levels increased 3 fold within 2 h (Fig. 8.8). In *Platymonas*, another wall-less alga, the same function is performed by mannitol, which accounts for 45% of the total osmotically active cellular solute 2 h after hyperosmotic shock.

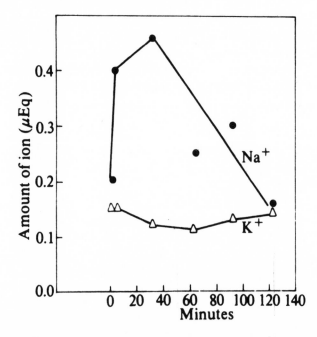

Figure 8.7. Changes of Na^+ and K^+ concentrations in *Dunaliella* following hypertonic stress. Note the initial increase in $[Na^+]$ which is restored to its original concentration within about two hours. (From Ginzburg, *Journal of Experimental Botany*, 32: 333-340, 1981)

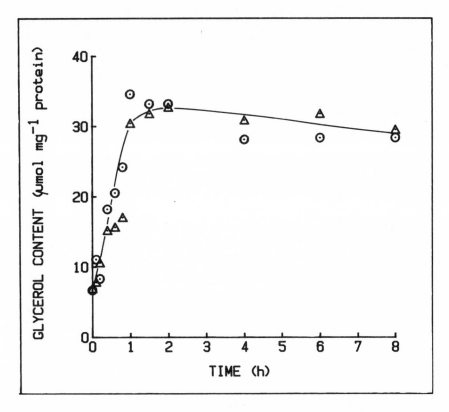

Figure 8.8. Increases in glycerol content of *Dunaliella* following transfer from 1.5 M NaCl to 4.0 M NaCl. (Unpublished data of D.S. Kessly, cited by Brown and Borowitzka in *Biochemistry and Physiology of Protozoa*, Vol. I: 139-190, M. Levandowsky and S.H Hutner, eds. Academic Press, 1979)

WALLED CELLS

Once a cell wall had evolved, volume regulation and the problems associated with lysis in dilute environments (e.g. fresh water) were eliminated. High hydrostatic pressures (turgor), associated with water entry into the cell, could be sustained due to the strength of the cell wall. Indeed, in walled cells turgor represents the physical driving force for cell expansion and plant growth. In addition higher plants have made use of changes in turgor for stomatal control, leaf movements (e.g. the sensitive mimosa), twining of tendrils, closure of Venus fly-traps and many other functions. If turgor falls below a critical value, due to inadequate supplies of soil water or as a result of osmotic effects caused by high external salt concentrations, then cell growth ceases.

For halophytes, those plants which can tolerate high concentrations of salts such as NaCl, Na_2SO_4, or $MgSO_4$ in their environments, the osmotic problem is considerable. For example, *Atriplex vesicaria* (Fig. 8.17) can tolerate NaCl concentrations up to 700 mM, while *Salicornia europa* is able to survive 1M NaCl. *Dunaliella*, has been reported to cope with 4M NaCl!

In order to maintain equilibrium it is evident that internal solute concentrations must be at least as high as those of the external medium, otherwise loss of turgor would occur. Among the halophytes high concentrations of K^+, Na^+ and Cl^- are accumulated to balance the high external salt levels. For example, in *Salicornia rubra*, a saltmarsh inhabitant, up to 90% of the total osmotic potential of the tissue may be accounted for by NaCl. Table 8.5 lists the major inorganic ions contained within the tissues of various halophytes found in the salt deserts of Utah.

Non-halophytes may be largely free of the salt problem (however, see Chapter 9), but they still face the difficulty of maintaining turgor when water is in short supply. Most higher plants generate turgor by means of the K salts of organic acids (e.g. malic acid). However, when K^+ is in short supply sugars may replace the turgor function of this ion (as in low-salt roots, Section 7.1) and Na^+ or other ions may compensate for lack of K^+ (Table 8.3).

TURGOR REGULATION

Both halophytes and non-halophytes appear to keep turgor pressure constant despite environmental fluctuations of salt concentration

Table 8.5. Inorganic ion concentrations (mM) in tissues of various halophytes. Plants listed were collected from the salt deserts of Utah. *Helianthus annuus* (the sunflower), a non-halophyte, is included for comparison. (Data from Wiebe and Walter, *The American Midland Naturalist*, 87: 241-245, 1972)

Plant	Na^{2+}	K^+	Ca^{2+}	Mg^{2+}	Cl^-	SO_4^{2+}
Helianthus annuus	36	192	12	11	101	19
Distichlis spicata	410	205	251	21	364	16
Salicornia europaea	576	172	0.3	32	535	17
Kochia americana	585	127	0.1	51	160	13
Salicornia utahensis	718	168	1.4	51	606	46
Atriplex confertifolia	858	409	0.2	56	666	54
Allenrolfea occidentalis	1044	173	0.2	23	526	104
Sueda depressa	1093	243	0.1	24	517	68

or water supply. The most elegant studies of turgor regulation have been conducted using giant algal cells like *Valonia* (Fig. 8.4c). When turgor pressure is reduced in these cells K^+ influx increases, and efflux decreases so that internal solute concentration increases, more water enters the cell and turgor is restored. Turgor sensing in these cells is thought to reside in the cell membrane but the mechanism(s) whereby turgor changes are transduced into counteractive responses are largely a mystery. In addition to the use of inorganic ions (K^+, Na^+ and Cl^-) halophytic higher plants also synthesize organic compounds (sugars, amino acids and other nitrogen compounds) to regulate turgor. These compounds are thought to be localized within the cytoplasm.

Many non-halophytes also maintain constant turgor pressure even as soil water disappears during drought. This process of "osmotic adjustment" has been studied extensively in crop plants. As stated in Section 7.5, when soils dry out the availability of inorganic ions tends to decrease. Not surprisingly, then, non-halophytes have evolved mechanisms for maintaining turgor (during drought) which are largely independent of external sources of solute. Indeed, most attempts to demonstrate increased K^+ uptake by roots which have suffered turgor reduction in laboratory experiments, have met with little success. In my own laboratory I have repeatedly attempted to demonstrate increased fluxes of K^+ in response to mild turgor reduction in barley roots. To no avail! The non-halophytic system for turgor regulation appears to depend upon internal resources for osmotic adjustment. Thus, following imposition of drought or osmotic stress, there are substantial changes of organic solutes, such as sugars and amino acids which bring about the necessary osmotic adjustment. Figure 8.9 illustrates the changes of sucrose and free amino acid concentrations in soybean seedlings which were transplanted to media containing $\frac{1}{8}$ (0.13X) the water provided to control plants (1.0X). It is very evident that solute content rose dramatically in the water stressed plants in the hours following imposition of this stress.

8.2b METABOLIC FUNCTIONS OF THE ESSENTIAL ELEMENTS

Traditionally, the elements have been divided into two groups, the macronutrients and micronutrients, according to their relative abundance in plant tissues. This system is convenient and widely accepted; its major weakness is that it provides little information about the members of each group. Table 8.6 provides a more functional classification, in which the elements are divided into 4 major groups:

Group 1 : The Structural Elements
Group 2 : The Enzyme Activators
Group 3 : The Redox Reagents, and
Group 4 : Elements of Uncertain Function.

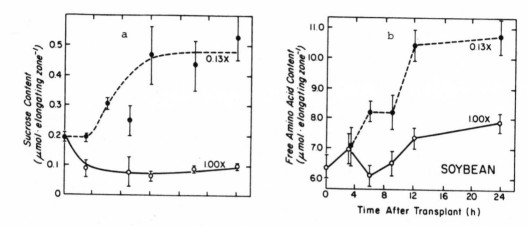

Figure 8.9.   Changes of solute contents of soybean seedlings which were transplanted to media containing ⅛ (0.13X) the water provided to control (1.0X) plants. (a) sucrose content, and (b) free amino acid content of the elogating zones (control plants: o; water stressed plants: •). (From Meyer and Boyer, *Planta*, 151: 482-489, 1981)

Table 8.6.   A list of the essential elements with typical concentrations (as % dry weight) in higher plants. Also included are Group and specific functions of the elements.

Group 1: Structural Components of Biological Compounds (carbohydrates, proteins, lipids, nucleic acids) and Intermediates of Metabolism		
Element	*Percent of Total Dry Weight*	*Major Functions*
C	44	Components of organic compounds
H	6	Components of organic compounds
O	44	Components of organic compounds
N	2	Amino acids, proteins, coenzymes, nucleic acids
S	0.5	Sulfur amino acids, proteins, coenzyme A
P	0.4	ATP, NADP, intermediates of metabolism (e.g. sugar phosphates) membrane phospholipids and nucleic acids

Group 2: Enzyme Activators; Elements Required for the Activation of Specific Enzymes

Element	Percent of Total Dry Weight	Major Functions
K	2.0	Activation of $\sim$ 60 enzymes (e.g. pyruvate kinase). Essential for protein synthesis. Responsible for turgor and turgor-based movements of guard cells and leaves.
Ca	1.5	Enzyme activator (e.g. α-amylase and membrane ATPases). Essential for membrane permeability. Associated with pectins in cell walls.
Mg	0.4	Activator of numerous enzymes, particularly ATP: phosphotransferases. A component of chlorophyll.
Mn	0.4	Activator of IAA oxidase, malic enzyme, and isocitrate dehydrogenase. Essential for photolysis of H_2O (possibly via redox changes, $Mn^{2+} \rightarrow Mn^{3+} + e^-$)

Group 3: Redox Reagents; Elements that Undergo Reduction/Oxidation (redox) Reactions by Virtue of Multiple Valency States

Element	Percent of Total Dry Weight	Major Functions
Fe	0.015	participant in cytochromes ($Fe^{III} + e^- \rightleftharpoons Fe^{II}$)
Cu	0.002	participant in cytochrome oxidase and plastocyanin ($Cu^{II} + e^- \rightleftharpoons Cu^{I}$)
Mo	0.002	Reduction of NO_3^- by nitrate reductase, and reduction of N_2 by nitrogenase of free living and nodule bacteria ($Mo^{VI} + e^- \rightleftharpoons Mo^{V}$)

Group 4: Elements of Uncertain Function

Element	Percent of Total Dry Weight	Major Functions
B	0.003	Thought to be important for membrane activity. Root growth very susceptible to deficiency.
Cl	0.01-2.0	Osmosis, charge balance, photolysis of water.
Si	<20	May serve as a strengthening agent in *Equisetum* and the grasses. Also thought to reduce transpiration.
Na	0.50-10.0	Requirement for plants using C_4 photosynthesis (including CAM plants). Source of turgor for halophytes.

Some of the elements have properties which could justify their presence in more than one group but I have used what is considered to be the main function to decide where to place such elements.

8.2c SPECIAL ELEMENTS FOR SPECIAL FUNCTIONS

The Structural Elements C, H, O, N, P and S

We have seen (Table 8.6) that C, H, O and N make up more than 96% of plant dry weight. The Periodic Table (Fig. 8.1) reveals that these elements (C, H, O and N) occupy the first two rows of the Table. These elements have small atomic numbers, their nuclei exert strong forces of attraction on valency electrons (because of the close proximity between electrons and nuclei) and atomic radii are small (Table 8.7). Moving across the second and third rows of the Periodic Table, as nuclear charge increases, the forces of attraction exerted on valency electrons tend to increase. This has two consequences: the energy required to remove an electron (the ionization energy) tends to increase, and atomic radii tend to decrease with increasing atomic number within a row. Thus in the first and second rows, F, N, O, H and C have the highest ionization energies and smallest atomic radii, while in the third row Cl, P and S share these attributes (Table 8.7).

Table 8.7: Atomic numbers, Atomic Radii (A°) and Ionization Energies (kJ mol^{-1}) for selected elements.

Element	Atomic Numbers	Atomic radii	Ionization Energy
H	1	0.4	1,312
Li	3	1.35	520
Be	4	0.90	900
B	5	0.80	800
C	6	0.77	1,086
N	7	0.70	1,402
O	8	0.66	1,314
F	9	0.64	1,681
Na	11	1.6	496
Mg	12	1.30	737
Al	13	1.43	577
Si	14	1.20	786
P	15	1.10	1,012
S	16	1.04	1,000
Cl	17	0.99	1,255

Covalent bonds are formed between elements when electron orbitals of separate atoms overlap to form a common bond orbital, shared by both electrons. Generally, the greater the orbital overlap, the stronger (more stable) the resulting bond, and as a rule the smaller atoms achieve greater orbital overlap than the larger atoms. In the cases of F and Cl, valency electrons are so tightly held that these elements tend to form ionic rather than covalent bonds. However, C, H, O, N and to a lesser extent P and S form the most stable of covalent bonds.

According to Wald (1962) another important property of the second row elements (particularly C, N and O), and to a lesser extent P and S in the third row is their capacity for multiple (double, triple) bond formation. The properties of many biologically important molecules can be ascribed to the presence of such bonds. For example the alternating double bonds in chlorophylls and carotenoids are responsible for their characteristic photochemical properties.

ENZYME ACTIVATORS: K^+, Ca^{2+}, Mg^{2+}, Mn^{2+} and Zn^{2+}

These are all metallic ions, which by virtue of their capacity to achieve stable (Inert Gas) electronic configuration through electron loss, participate in ionic bonds. In solution these charged species (K^+, Mg^{2+}, etc.) bring about pronounced changes in the physical properties of the enzymes with which they become associated. For example, experiments with pyruvate kinase (a K^+-requiring enzyme) and antibodies prepared against this enzyme indicate that marked conformational changes were associated with K^+ binding to the enzyme. Control experiments with catalase (an enzyme with no K^+ requirement) gave no indication of such changes. Similar conclusions are reached from kinetic studies, which indicated that the univalent activators may exert allosteric effects on enzymes such as K^+-activated ATPases. However, as yet the precise mechanisms responsible for these effects are not fully understood.

REDOX REAGENTS: Fe^{3+}, Cu^{2+} and Mo^{5+}

The energy which is required for the many biological processes in which living organisms engage is derived from respiration (the oxidation of food materials) or photosynthesis. It is evident that the major reduction/oxidation pathways (mitochondrial and chloroplast electron transport), bring about a gradual 'step by step' release of energy. The free energy changes associated with these reactions is released in small measures through coupled reactions of the sort shown below:

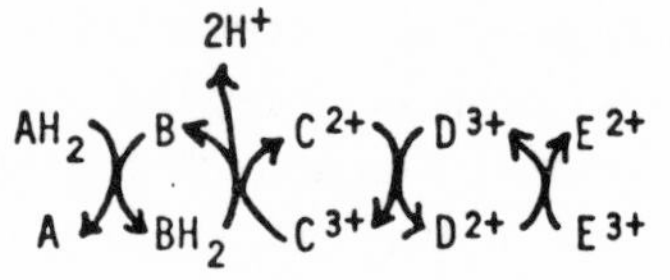

$$AH_2 \underset{A}{\overset{B}{\bigtimes}} \underset{BH_2}{\overset{2H^+}{\bigtimes}} \underset{C^{3+}}{\overset{C^{2+}}{\bigtimes}} \underset{D^{2+}}{\overset{D^{3+}}{\bigtimes}} \underset{E^{3+}}{\overset{E^{2+}}{\bigtimes}}$$

The reduced compount AH_2 is capable of reducing B and in the process is itself oxidized. It transfers H^+ and electrons to B and increases the free energy of B, while its own free energy is diminished. Like all spontaneous processes, AH_2 can readily reduce B because it has a higher free energy (more negative redox potential) than BH_2. The reverse reaction is unlikely on thermodynamic grounds. The next step, in which BH_2 reduces C^{3+} to C^{2+}, involves the transfer of an electron to C^{3+}, but not the H^+. It is evident that the participants in such a sequence must possess the capacity for reduction/oxidation (redox) reactions.

Redox reactants all contain metallic ions (Fe^{3+}, Cu^{2+} or Mo^{5+}) which can reversibly undergo redox reactions:

$$Fe^{III} + e^- \rightleftharpoons Fe^{II}$$

The complete redox agent, e.g. cytochrome b, also consists of the enzyme protein, with a firmly bound organic cofactor (or prosthetic group), which is the porphyrin molecule (Fig. 8.10). At the centre of the porphyrin molecule is an atom of Fe.

In order to bring about the sequential oxidations referred to above it is necessary to have many redox reagents, each with different redox potentials (see Section 3.5). The redox potential for the redox

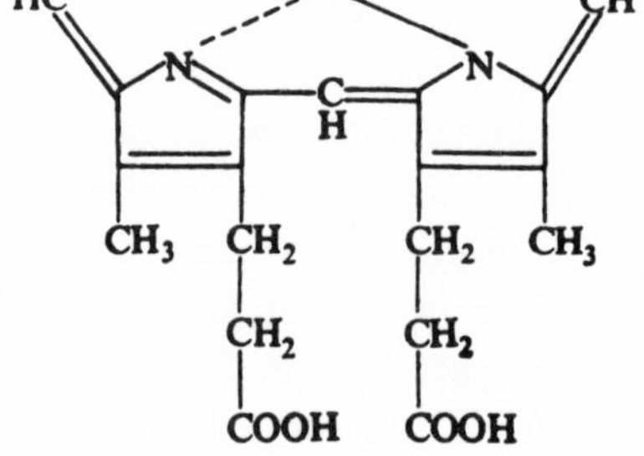

Figure 8.10.   The porphyrin molecule of a typical cytochrome.

reagent is influenced, not only by the identity of the metal (Fe, Cu, Mo) which is associated with it, but also by the nature of the prosthetic group and its interactions (bonds) with the enzyme protein.

8.2d   THE FUNCTIONS OF GROUP 1 ELEMENTS

CARBON, HYDROGEN AND OXYGEN:

Of the 30 billion tons of carbon fixed annually by higher plants each year, about one third is converted to cellulose (a polymer of glucose, Fig. 8.11). Massive amounts of lignin, an aromatic derivative, are also formed, particularly in forest trees. In addition the many small molecules such as sugar phosphates, amino acids and fatty acids which participate in intermediary metabolims are made up of these elements (see Fig. 8.11).

NITROGEN:

The bulk of organic nitrogen occurs as protein, nucleic acids and amino acids. Other important nitrogen compounds are the purines, e.g. adenine which forms part of the structure of coenzymes such as ATP and NAD^+, and which, together with the pyrimidines (also N compounds) form the organic bases of the nucleic acids (Fig. 8.12). In addition to the twenty amino acids which make up proteins there are many 'non-protein' amino acid derivatives, some of which are

Figure 8.11.   Representative examples of the major classes of organic compounds

(a) Carbohydrates *(C, H, O)

Glucose: a monosaccharide

$(C_6H_{12}O_6)$

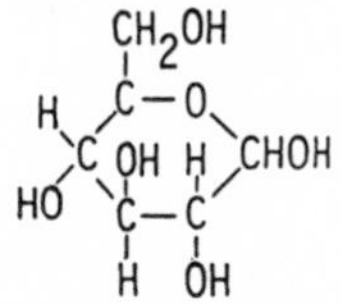

*Most sugars participate in metabolism as the phosphate esters

(b) Proteins †(C, H, O, N)

Alanine: an amino acid

$(C_3H_7O_2N)$

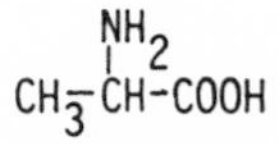

$$CH_3-CH-COOH$$
$$\overset{NH_2}{|}$$

†The amino acids cysteine, cystine and methionine also contain S

(c) Lipids ♦(C, H, O)

Oleic acid: an unsaturated plant oil

$(C_{18}H_{34}O_2)$

$$CH_3(CH_2)_7CH=CH(CH_2)_7COOH$$

♦Phospholipids (membrane constituents) also contain P

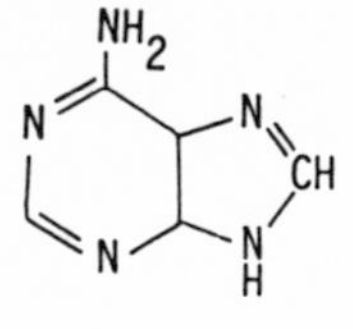

Adenine (a purine)

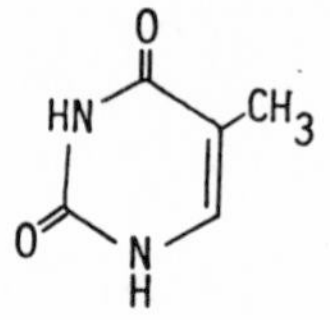

Thymine (a pyrimidine)

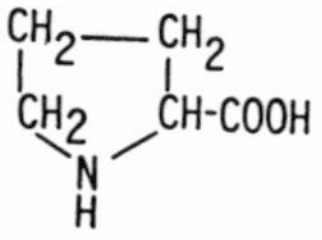

Proline (an imino acid)

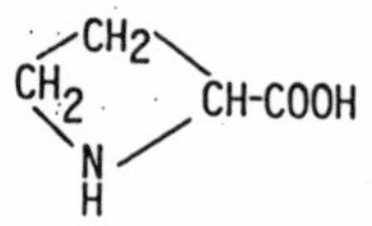

Azetidine-2-carboxylic acid

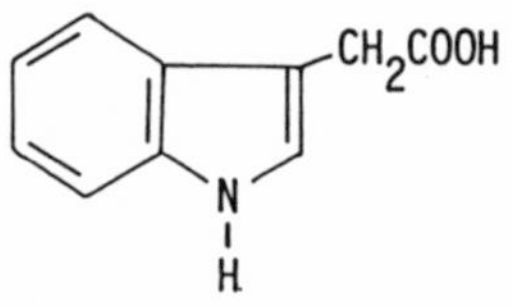

Indole Acetic Acid

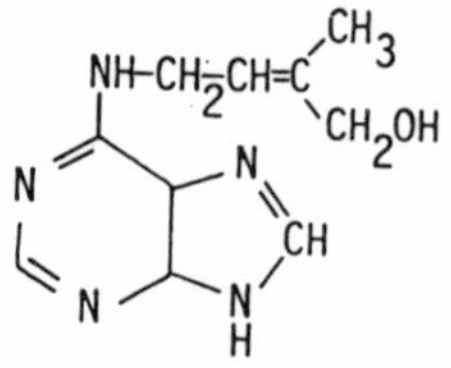

Zeatin (a cytokinin)*

*note similarity to adenine

Figure 8.12.   Structure of Some Nitrogen Derivatives

extremely toxic. For example, members of the Lily family such as Lily of the Valley contain considerable amounts of azetidine carboxylic acid (Fig. 8.12), a close structural analog of the imino acid proline. This compound proves toxic when administered to plants or animals because it is incorporated into proteins in place of proline. Interestingly, species which possess this compound naturally never confuse it with proline.

The plant growth regulators, indole acetic acid and cytokinins (Fig. 8.12) as well as numerous alkaloids and other pharmacologically active compounds also contain N.

SULFUR:

Like nitrogen, sulfur also participates in protein structure, but as part of the S-amino acids cysteine and methionine. The formation of disulfide linkages (S-S) between adjacent cysteine residues is responsible for linking and stabilizing polypeptide structure (Fig. 8.13).

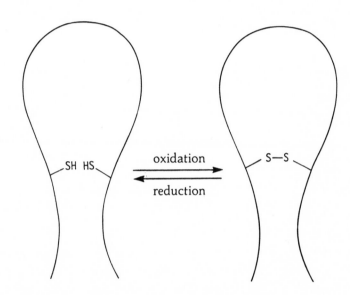

Figure 8.13.   Stabilization of polypeptide structure through formation of disulfide linkages between adjacent cysteine residues.

The enzyme ribonuclease (Fig. 8.14) contains a total of 8 cysteine residues at amino acid positions 26, 40, 58, 65, 72, 84, 95 and 109. These are involved in 4 disulfide linkages. When these bonds are broken by reducing agents the conformation of the protein is disrupted and enzyme activity is lost. Removal of the reducing agent causes gradual spontaneous reoxidation of the sulfhydryl groups and recovery of enzyme activity.

In addition to serving a structural role, the SH groups of these protein amino acids participate directly in enzyme-catalyzed reactions. Coenzyme A contains a SH group which can form a thio (sulfur) linkage (S-C) with molecules such as the fatty acids. Linkage to CoA activates such molecules which can then participate in the TCA cycle and the synthesis and degradation (β-oxidation) of fatty acids. Coenzyme A also participates in the biogenesis of a wide array

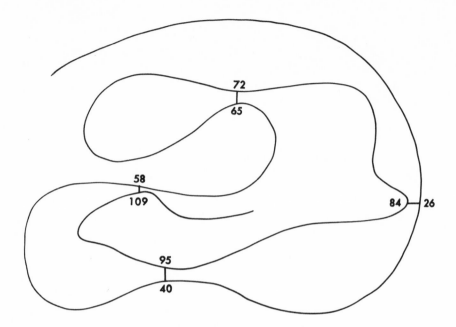

Figure 8.14. Location of 4 disulfide linkages in the enzyme ribonuclease between amino acids at positions 26 and 84, 40 and 95, 58 and 109 and 65 and 72.

of important terpenoid compounds, including many of the plant perfumes, gibberellins, carotenoids and steroids. Some rather unique, often strong tasting and evil smelling, sulfur derivatives are to be found among members of the cabbage family (e.g., mustard) as well as in the onion family (such as garlic!).

PHOSPHORUS:

This element is principally available to plants as the orthophosphate (Fig. 8.15). Unlike NO_3^- and SO_4^-, however, P serves its biological functions as esters (anhydrides) with various organic molecules including sugars, carboxylic acids and glycerol. Best known of the phosphate anydrides are those produced by covalent bond formation with other phosphate groups as in ADP, ATP and certain polyphosphates. Another important P derivative is phytic acid, (Fig. 8.15) the hexaphosphate of inositol, which is an important seed reserve of P.

The principal role of P is in energy transduction. During electron flow along the electron transport chains of mitochondria and chloroplasts, the free energy changes associated with the oxidation of the chain's components is conserved in ATP. It is now widely held that the redox steps of the electron transport pathway are linked to vectorial

Orthophosphate (in ionized form)

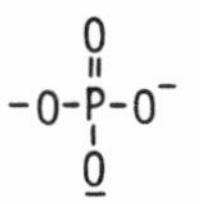

$$-O-\overset{\overset{\displaystyle O}{\|}}{\underset{\underset{\displaystyle O}{|}}{P}}-O^-$$

Phytic Acid
(The hexaphosphate of myo-inositol)

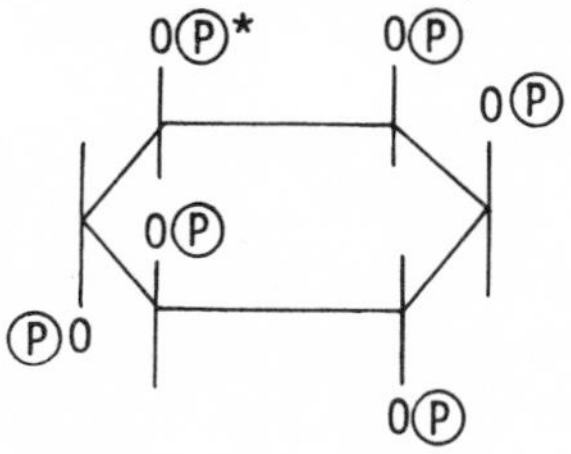

*(P) = orthophosphate

Nucleic Acids (e.g. RNA)

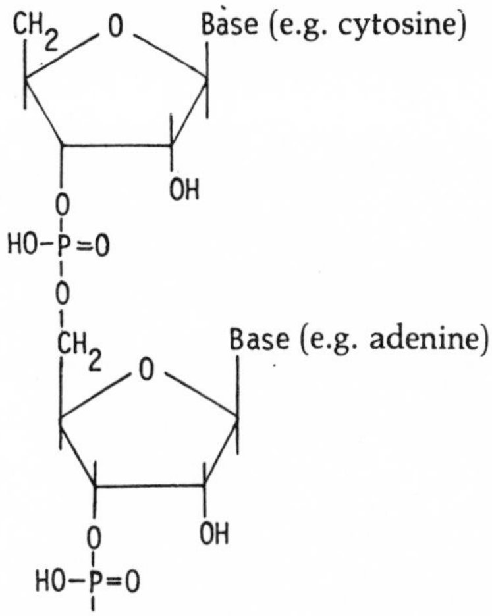

Base (e.g. cytosine)

Base (e.g. adenine)

Lecithin (a phospholipid)

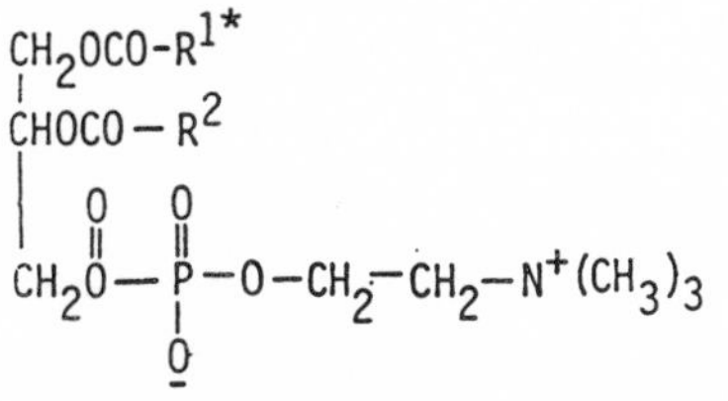

*R^1 and R^2 stand for fatty acids such as oleic acid

Figure 8.15.   Structural formulae of some important phosphate derivatives.

H^+ transport across the thylakoid and cristae membranes. As a consequence, during electron flow, pH and electrical gradients are set up across the transducing membranes. These gradients serve as the energy source for the synthesis of ATP (Section 5.7). The formation of ATP during electron transport in chloroplasts represents the primary biochemical energy transduction upon which all subsequent metabolic reactions depend. Orthophosphate also serves a linking function as in sugar-phosphate-sugar chains of nucleic acids and glycerol phosphates of membrane phospholipids (Fig. 8.15). In the latter group of compounds the Pi moiety provides a polar end to the phospholipid which is critical for membrane function (see Section 5.3).

8.2e   THE FUNCTIONS OF GROUP 2 ELEMENTS

Some enzymes consist solely of polypeptides and require only the appropriate substrate for activity. In other cases, however, enzymes may also require inorganic cofactors such as K^+, Ca^{2+}, Mg^{2+}, Mn^{2+} and Zn^{2+}. The association between inorganic cofactor and the enzyme is essential for normal activity. In the absence of the required inorganic ion, enzyme activity may be reduced to zero. The enzyme activator may provide an appropriate ionic environment so that the necessary enzyme conformation is maintained. Such is the case for many K^+-activated enzymes. In other cases e.g., Mg^{2+}, the enzymes may require the ion to form a link with the substrate.

POTASSIUM:

At least sixty enzymes are known to be activated by this ion. The enzyme pyruvate kinase (more correctly referred to as ATP: pyruvate phosphotransferase), which participates in glycolysis, is perhaps one of the best studied of the K^+-activated systems:

In addition to K^+, this enzyme also requires Mg^{2+}. The latter ion is believed to participate in an enzyme-ADP-Mg complex which is an intermediate in the formation of ATP:

Potassium is thought to be essential for the conformational stability of this enzyme. *In vitro* studies in our laboratory indicate that pyruvate kinase activity is extremely low when K^+ is omitted from the reaction mixture (Table 8.8 and Fig. 9.8). Figure 9.8 shows that pyruvate kinase activity appears to plateau at about 20 mM K^+. However, in the same variety of barley, we have estimated cytoplasmic $[K^+]$ to be about 150 mM. This high concentration of K^+ is thought to be essential for normal protein synthesis. Potassium's role in this process is considered to be the maintenance of a proper association between tRNA molecules and ribosomes during the translation of mRNA.

As discussed in Section 8.2a on Walled Cells, K^+ serves a critical role in the turgor-mediated opening of guard cells. Stomatal aperture, which is determined by guard cell turgor, is the principal means of water vapour and gas exchange between leaves and air in land plants. Increased turgor of guard cells during stomatal opening is brought about by increased absorption of K^+ into these cells. During stomatal closure there is a loss of K^+ from the same cells. Figure 8.16 shows these changes in guard cells of *Phaseolus*.

MAGNESIUM:

Magnesium is well known for its participation in the chlorophyll molecule, however, the major part of cell Mg (70% or more) is freely diffusible within the cytoplasm. Plant tissues contain Mg^{2+} in association with both mobile anions such as malate and citrate and with non-diffusible (insoluble) anions such as oxalate and pectate. Much of cellular ATP and ADP is also complexed with Mg^{2+}. In the seeds of

Table 8.8.   Pyruvate kinase activity of enzyme extracts from barley leaves (μmol pyruvate formed g^{-1} leaf h^{-1}). Enzyme incubation medium contained 10 mM $MgSO_4$, 2.5 mM ADP, 1.5 mM phosphoenolpyruvate and various concentrations of KCl. (From Memon and Glass, *Journal of Experimental Botany*, 36: 79-90, 1985)

Incubation Medium	Enzyme Activity
	($\mu mol\ pyruvate\ g^{-1}h^{-1}$)
0 mM KCl	0.06
5 mM KCl	15.72
50 mM KCl	20.72

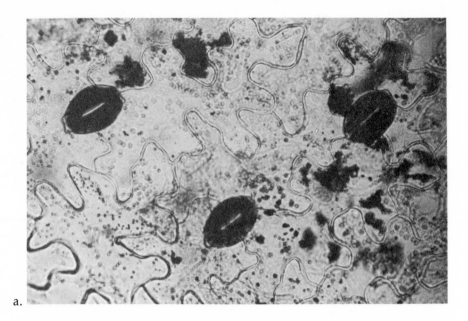

a.

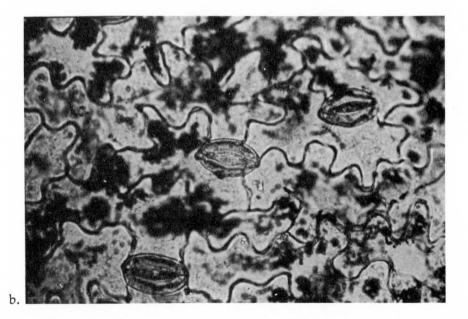

b.

Figure 8.16. Epidermal cells of Phaseolus leaves stained to show K$^+$ content. In 7.15a stomata are open and guard cells show strong staining reaction for K$^+$. In 7.15b stomata are closed and guard cells show a weak staining reaction for K$^+$. (Micrographs courtesy of Dr. T.A. Mansfield)

cereal grains Mg is present as the principal cation associated with phytic acid. Many enzymes, including phosphatases, and particularly those involving ATP, require Mg^{2+} for their activity. As discussed in the preceding paragraph, Mg ions participate in the enzyme -ATP-Mg complex involved in ATP-mediated reactions.

In many cases the enzyme activating effects of Mg can be replaced *in vitro* by Mn^{2+}, although the much lower tissue concentrations of Mn^{2+}, compared to Mg^{2+} (estimated at .3 and 16 mM respectively) raises doubt as to the physiological significance of this substitution.

In chloroplasts the light-driven pumping of H^+ into the thylakoid spaces, associated with electron flow, is accompanied by release of Mg^{2+} from thylakoids so that stromal Mg concentration rises. The opposite movements occur when plants are switched to darkness. Besides balancing the inward flux of H^+, this increase of stromal Mg may serve to activate the carbon fixing enzyme ribulose 1,5-bis-phosphate carboxylase, thus co-ordinating the light reactions and the carbon-fixing reactions.

CALCIUM:

It is generally considered that the principal functions of Ca^{2+} are performed extracellularly, in the cell wall and at the external surface of the plasmalemma. Because of its divalency and its thin hydration shell, Ca^{2+} complexes quite firmly to charged surfaces. In cell walls Ca^{2+} ions are found in association with the negatively charged carboxyl groups of the pectic fraction where their ability to cross link between different molecules is important in strong gel formation in the middle lamella. Particularly in soft fruits, Ca^{2+} serves to strengthen the cell walls and inadequate levels of this ion may lead to tissue swelling and cracking of the fruit. Calcium is also essential for normal cell membrane permeability. Membranes become leaky in the absence of external calcium and may lose their characteristic selectivity. As a consequence experimenters routinely include a Ca salt (e.g., 0.5 mM $CaSO_4$) in all uptake media. Calcium also appears to function intra-cellularly, and is involved in cell division and in the assembly of microtubules.

Despite the high concentration of Ca^{2+} in cell walls, and in some instances in the cell vacuole (where Ca^{2+} may be present as insoluble calcium oxalate), the concentration of this ion within the cytoplasm appears to be low ($\sim 10^{-7}M$) in all living organisms. This situation is achieved by extrusion mechanisms at the plasmalemma and the tonoplast together with the ability of mitochondria and other organelles (e.g., endoplasmic reticulum) to sequester Ca^{2+} from the cytoplasm.

In recent years animal physiologists have established that Ca^{2+}, acting together with cyclic AMP (cAMP), determines cellular responses to various external stimuli. In nerve cells and muscle cells, for instance, Ca^{2+} mediates between the electrical stimulus and subsequent reactions. Incubation of rat liver cells in abnormally low Ca^{2+} solutions leads to arrested DNA synthesis and cessation of cell division. When Ca^{2+} levels are restored DNA synthesis commences within 15 min. The discovery that plant cells contain cAMP and possess rather similar calcium-binding regulatory proteins to those found in animals (the Calmodulin proteins) has been interpreted to indicate that Ca^{2+} may act (together with Calmodulin) to control various cellular processes in plants. For example, three plant enzymes, namely, NAD kinase, Ca^{2+}–ATPases (which are associated with many membrane fractions) and membrane-bound protein kinases (enzymes responsible for phosphorylating membrane proteins) are thought to be activated by a calcium: calmodulin complex. According to some authors, e.g. J.B. Hanson (1984), calcium may represent *the* regulatory cation of eukaryotic organisms with a fundamental role in many aspects of plant development.

MANGANESE:

Although, Mn^{2+} may substitute for Mg^{2+} in some enzyme reactions *in vitro*, and may even be more effective in some cases, it is not clear that Mn^{2+} is the *physiological* activator even in the latter cases. However its ability to bind firmly to proteins, causing conformational changes in the latter, is more likely to be the characteristic which makes this element essential. In the photolysis of water associated with photosystem II, a requirement for four Mn per P680 reaction centre has been established. The capacity for redox changes, Mn^{3+} (manganic) to Mn^{2+} (manganous), may be crucial to the photo-oxidation of $2H_2O$ to yield O_2, but this point remains equivocal. In this regard, it is interesting that Mn deficiency in dicots is characterized by small yellow (chlorotic) spots on younger leaves. The enzyme indole acetic acid oxidase (IAA oxidase), which is thought to bring about destruction of IAA, appears to be activated specifically by Mn^{2+}. This may account for the fact that Mn deficiency is associated with high IAA activity in cotton plants, and Mn toxicity has been reported to be accompanied by increased activity of IAA oxidase.

ZINC

This element is required in quite small amounts in plant tissues. Generally a concentration of around 0.01% of dry matter is sufficient for healthy growth. Zinc serves as a cofactor, forming stable complexes with many important enzymes including several dehydrogenases (e.g. lactic dehydrogenase), aldolases, phosphatases, DNA and RNA polymerases and the enzyme carbonic anhydrase which is localized within chloroplasts and catalyses the reaction between H_2O and CO_2 to form bicarbonate:

$$H_2O + CO_2 \xrightarrow{\text{carbonic anhydrase}} H^+ + HCO_3^-$$

The list of such metalloenzymes involving Zn now numbers over 80. The diverse functions of these enzymes make it difficult to be precise regarding the function of Zn in these enzyme-bound complexes.

In rural districts of Iran and Egypt where large quantities of cereal grains are consumed in the diet in the form of unleavened bread, human Zn deficiency has been reported. The strong binding of Zn to phytic acid is thought to make Zn unavailable for intestinal absorption. Where breads are made with yeasts the live yeasts hydrolyze the Pi groups from phytic acid and eliminate this deficiency problem.

8.2f   THE FUNCTIONS OF GROUP 3 ELEMENTS

IRON

Iron is absorbed by plants in the ferrous form (Fe^{2+}). Indeed, where Fe^{2+} iron is not readily available, some varieties of sunflower, (referred to as iron-efficient) are known to excrete reducing compounds which convert the ferric (Fe^{3+}) form to the ferrous form. Within the plant the biochemical functions of Fe depend upon its capacity for chelate formation and its ability to participate in redox reactions through the interchangeability of its valency states ($Fe^{3+} + e^- \longrightarrow Fe^{2+}$). This function is principally performed within the porphyrin molecule (Fig. 8.10) in which an atom of Fe is bound to the N atoms of the 4 pyrrole groups of the porphyrin (like Mg^{2+} in chlorophyll). Such complexes form the prosthetic groups of several important enzymes, including the cytochromes and peroxidase. In addition, Fe is present in a group of iron-sulfur enzymes which have no heme (iron-porphyrin) structure. One of the best known of these is ferredoxin.

COPPER

This element also exists in two valency states, namely cuprous (Cu^+) and cupric (Cu^{2+}), although the latter is the form which is chiefly available to plants. Generally, Cu is required in only small amounts and high concentrations are potentially toxic. Like iron, this element participates in redox reactions, of which the best known are probably those involving reaction with oxygen. Cytochrome oxidase e.g., which reduces molecular O_2 to H_2O in the last step of the respiratory electron transport chain contains this element. However, in leaf tissue Cu is mainly found within the chloroplasts in the redox enzyme plastocyanin which links photosystems I and II. In addition, other important Cu enzymes are superoxide dismutase, which protects aerobic organisms from the potentially harmful superoxide radical O_2^- (obligate anaerobes lack this enzyme and cannot survive in the presence of O_2), and phenol oxidase, the enzyme responsible for hydroxylating and oxidising phenols in lignin biosynthesis.

MOLYBDENUM

This element is available to plants as molybdate (MoO_4^{2-}). Although it is required in very small quantities (0.0001% of dry weight is sufficient), its participation as a component of the enzyme nitrate reductase is vital to most plants. The source of reducing power for the reduction of NO_3^- comes from the coenzymes NADH or NADPH which in turn reduce FAD and Mo. The valency change Mo^{5+} to Mo^{6+} is probably responsible for the final reduction of NO_3^- to NO_2^- :

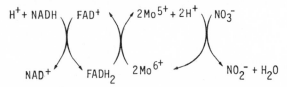

The element is also a participant in the nitrogenase enzyme of root-nodule bacteria responsible for the reduction of N_2 in the roots of legumes.

8.2g   GROUP 4: ELEMENTS OF UNCERTAIN FUNCTION

BORON

Boron occurs in soil mainly as borate (BO_3^-) and it is in this form that it participates in plant nutrition. However, animals, fungi and microorganisms appear not to require this element so that its function must be related to some unique feature of plant metabolism. Thus, although B has been implicated in the synthesis of nucleic acids, based upon reduced RNA content and reduced incorporation of ^{32}P into DNA and RNA associated with B deficiency, it is difficult to establish the direct connection between B and these processes which is unique to plants. The rapid recovery of membrane transport when B is added to B-deficient roots has led to speculation that B acts upon some membrane process. If this is the case then it is clear that all manner of subsequent events might be indirectly influenced. Root tips appear to be particularly sensitive to the removal of B from culture media. Recent work with sunflowers has demonstrated that root elongation is inhibited within 3 h of removal of this element.

Electron microscope studies reveal that by 6 h there are unusual cell wall thickenings and indications of a loss of membrane integrity.

CHLORINE

Chloride is accumulated in quite large amounts by plants in the field although under experimental conditions its absorption and accumulation can be almost completely suppressed when nitrate is available. A definite role for Cl in photosynthetic oxygen evolution has been proposed on the basis of its requirement in isolated chloroplasts. However, the *in vivo* requirement for this function has been questioned. It is probably as a counter-ion for K^+ that Cl^- is required in significant quantities, but in this role it is subservient to NO_3^-, the 'preferred' anion. It is important to remember that up to 98% of the potassium used in potash fertilizer may be in the form of KCl.

SILICON

Below pH 9, Si is available to plants as the monosilicic acid $Si(OH)_4$. Above this pH it occurs as the silicate ion. Silicon is found in the tissues of many plant species but it is particularly important in the grasses (especially rice) as well as in the sedges, horsetails (*Equisetum* sp.) and in diatoms. In rice and in *Equisetum*, Si may amount to 20% of the plant dry weight. Within plant tissues Si occurs as the hydrated amorphous silica (silica gel). The polymerization of silicic acid to form silica gel is pH dependent and the reaction is non-reversible so that when formed in a particular location the deposit is permanent. Silica gel readily complexes with proteins and carbohydrates such as cellulose so that its deposition may involve chemical linkage with the cell wall polysaccharides. Silicon is deposited within the cell lumen in leaves of forest trees and the secondary wood of perennial species. It is deposited in cell walls, particularly in epidermal cells and in epidermal hairs in intimate association with cellulose. In addition, scanning electron micrographs of stems of *Equisetum* and leaves of rice and oat reveal numerous silica knobs distributed over the leaf surface beneath the cuticle (see Fig. 8.17). In higher plants Si is considered to add strength to the cell walls in which it is deposited, as well as reducing water loss from leaves and affording some protection against fungal infection.

SODIUM:

Sodium is now considered to be an essential nutrient for plants capable of fixing CO_2 via C_4 organic acids. This includes C_4 photosynthetic species, e.g. *Atriplex vesicaria* (Fig. 8.18), and plants which use the Crassulacean Acid Metabolism (CAM plants). In addition certain blue-green algae (Cyanophyta) and marine bacteria and fungi

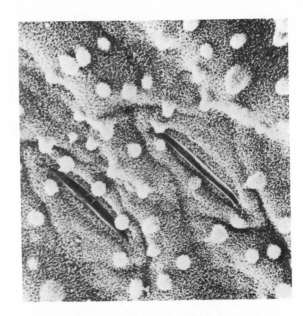

Figure 8.17.   Scanning electron micrograph of silica deposits on leaves of rice. (From Troughton and Sampson, *Plants: A scanning electron microscope survey*, John Wiley and Sons, 1973)

Figure 8.18.   Sodium deficiency symptoms in *Atriplex vesicaria*. (From Brownell, *Plant Physiology*, 40: 460-468, 1965)

have been shown to require this element. Halophytes characteristically accumulate high concentrations of Na in order to balance the low $\Psi\pi$ of the external environment. However the element is probably localized within vacuoles because enzymes from halophytic species seem to be as sensitive to salt as those from nonhalophytes. Brownell has listed several categories of plants with respect to sodium:

1. Species for which Na is an essential nutrient, such as *Atriplex vesicaria*.

2. Species which demonstrate beneficial responses to relatively high Na application even when all other required elements are available. Members of the Chenopodiaceae, e.g. beet (*Beta vulgaris*), many of which are restricted to salty habitats, are in this category.

3. Species which respond to high levels of Na application only when K is limiting growth, e.g. barley, wheat and oat or, in the case of cotton, when Ca is available.

4. Those species which demonstrate little or no response to sodium even when elements such as K are limiting growth, for example cucumber, potato and sunflower.

There is little clear understanding of the role of Na in those species which require this element. Its requirement by species which fix carbon in C_4 organic acids has led to the proposal that Na is necessary for the operation of the C_4 photosynthetic and CAM pathways. However this proposal is largely based on circumstantial evidence.

8.3   Special Topics

8.3a   REDUCTION AND ASSIMILATION OF NO_3^- AND SO_4^{2-}

Both N and S are absorbed from soil solution as the oxides NO_3^- and SO_4^{2-}, which must be reduced to NH_3 and S^-, respectively, before they can be incorporated into organic compounds. In many of the plant species which have been examined (including maize, barley, lupin and radish) NO_3^- reduction, by the enzyme nitrate reductase, occurs in roots as well as in shoots. Thus the xylem sap of these plants may contain significant quantities of reduced N in the form of amino acids (particularly nitrogen-rich amino acids glutamine and asparagine). The enzymes responsible for nitrate reduction, nitrate reductase (NR) and nitrite reductase (NiR) have been studied extensively. The induction of NR by NO_3^- and repression by NH_4^+ is a model system for the study of enzyme regulation.

Reactions of Nitrate Reduction

(a) $NO_3^- \xrightarrow{\quad\text{nitrate reductase}\quad} NO_2^-$

$$NADH + H^+ \qquad NAD^+ + H_2O$$

(b) $NO_2^- \xrightarrow{\quad\text{nitrite reductase}\quad} NH_3 + H_2O + OH^-$

$$6e^- + 6H^+$$

In leaves the enzyme NiR is located within the chloroplast and the electron donor is ferredoxin. In roots NiR occurs in proplastids but the electron source appears to be unknown.

When NO_3^- has been reduced to the level of NH_3 it is incorporated into the amino acid glutamine in the reactions of the GS–GOGAT pathway (Fig. 8.19). For each turn of the cycle 1 molecule of α ketoglutarate is consumed, a molecule of glutamate is

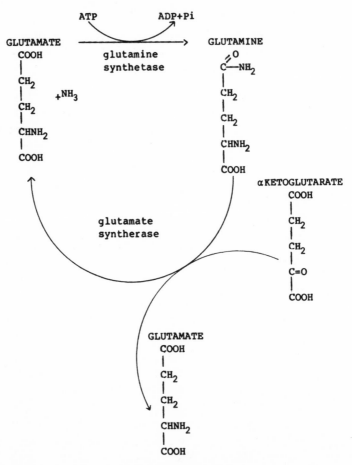

Figure 8.19.   Reactions of the GS–GOGAT Pathway

produced, and the original starter molecule of glutamate is regenerated. The amino group of glutamate can then be passed to various keto acids (the process of transamination) to generate other amino acids, e.g.

$$\text{glutamate} + \text{pyrurate} \longrightarrow \alpha \text{ ketoglutarate} + \text{alanine}$$

The reduction of SO_4^{2-} is preceded by its 'activation' in a reaction catalyzed by the enzyme ATP sulfurylase, in which SO_4^{2-} replaces the terminal phosphate groups of ATP to generate an adenosine monophosphosulfate derivative (APS) (Fig. 8.20).

The details of subsequent steps have been worked out for *Chlorella*, *Euglena*, spinach chloroplasts and for the duckweed (*Lemna*). The activated sulfate (APS) is reduced by ferredoxin following its transfer to an SH containing (thiol) carrier. In *Chlorella* this carrier is glutathione, a tripeptide of glutamic acid, cysteine and glycine, but in spinach a somewhat larger molecule is involved. The valence state of S in SO_4^{2-} is +6 and in reduced sulfur (S^{2-}) −2 so that 8 electrons are required to bring S to the appropriate valence state for incorporation into the S-amino acids. The transfer of reduced S to O-acetyl serine effectively replaces the acetate moiety of the latter molecule to generate cysteine. This entire pathway is shown in Figure 8.20.

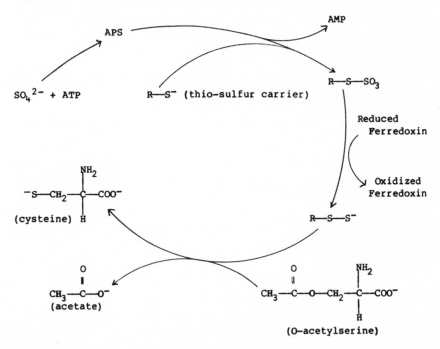

Figure 8.20.   Reactions of sulfate reduction.

8.4 Deficiency Symptoms

INTRODUCTION

The disruption of biochemical function arising from inadequate supplies of particular elements is translated at the whole plant level into diverse and sometimes highly characteristic deficiency symptons. These are extremely useful to the experienced plant nutritionist in diagnosing the deficiency problem. However the situation may be extremely complicated in the field due to the many sources of interactions with other environmental variables. For example, loss of green color (chlorosis) is a common symptom of Fe deficiency, which may occur even when the soil contains adequate Fe. One form of Fe-deficiency chlorosis, referred to as "lime-induced" chlorosis, is caused by liming of the soil which raises its pH. As a consequence of the low solubility product of Fe^{3+} salts, raising the pH one unit (from 6 to 7) reduces soluble Fe^{3+} concentration by a factor of 1000. Because the visual symptoms of deficiencies may be many steps removed from the impaired biochemical function, the symptoms of particular deficiencies may appear unrelated to the underlying causes. Consequently the symptoms of elemental deficiency tend to be rather empirical.

Furthermore, differences in the nature and the extent of deficiency symptoms can be quite dramatic among various species and even varieties of the same species grown under identical conditions. Thus, apart from some broad generalizations, symptoms of specific element deficiencies are often peculiar to particular species.

8.4a   THE DETAILS OF DEFICIENCY SYMPTOMS IN HIGHER PLANTS

The following is a brief outline of the major symptoms associated with deficiencies of the essential elements (see Fig. 8.2).

NITROGEN:

1. Stunted growth.
2. Chlorosis beginning in older leaves. Tips and margins of leaves commonly become yellow first.

SULFUR:

1. Plants tend to be small and spindly.
2. Young leaves chlorotic.
3. Delayed maturity.

PHOSPHOROUS:

1. Stunted growth.

2. Older leaves become dark green with purple coloration (due to anthocyanin development) in some species.

POTASSIUM:

1. Slow growth and susceptibility to wilting.
2. Dark green foliage with necrotic spots appearing on older foliage, typically at tips and margins of leaves.

MAGNESIUM:

1. Chlorosis between veins and along margins of older leaves.
2. Leaves tend to be brittle and may fall prematurely.

CALCIUM:

1. Reduced growth or even death of apical meristems, often leading to multiple branching in tap root crops.
2. Young foliage may be abnormal, chlorotic or even 'burned' at tips.
3. Softening of tissues and wall breakdown is common in fruits.

MANGANESE:

1. Interveinal chlorosis which shows up in younger leaves first.
2. Cereal leaves may develop grey specks (oats) white streaks (wheat) or brown spots and streaks (barley).

ZINC:

1. Reduced stem growth may cause "rosette" condition of terminal leaves.
2. Chlorosis often appears as yellow mottling, between the veins in younger leaves.

IRON:

1. Young leaves demonstrate interveinal chlorosis.
2. In cereals chlorosis may appear as yellow stripes, alternating with green, along the length of leaves.

COPPER:

1. Stunted growth and "dieback" of young twigs in fruit trees.
2. In cereals white tipped leaves are an early symptom.
3. Reproductive growth strongly affected so that fruit may not be formed in many species.

BORON:

1. Stunted growth or death of apical meristems, followed by sprouting of lateral buds.

2. Leaves may be thickened, curled and chlorotic.

3. Death of meristems causes reduced flower and fruit set.

CHLORINE:

1. Leaves chlorotic and susceptible to wilting.

2. Deficiency is practically nonexistent in nature because even rain water will carry sufficient Cl^- to provide the required amounts.

SILICON:

1. In rice and *Equisetum* a wilting growth habit may result.

2. Abnormal accumulation of Fe and Mn may cause necrotic spots.

Summary

The functions performed by the essential elements are both biophysical and biochemical in nature. The presence of inorganic ions, particularly K^+, is responsible for the generation of turgor, a prerequisite for plant growth and stomatal function. The particular biochemical functions of the 16 essential elements are determined by their unique chemical and physical properties. Inadequate supplies of even one of these elements lead to reduced growth and characteristic deficiency symptoms.

Further Reading

Clarkson, D.T. and Hanson, J.B. 1980. The mineral nutrition of higher plants. *Ann. Rev. Plant Physiol.* 31: 239-98.

Epstein, E. 1972. *Mineral Nutrition of Plants: principles and perspectives*. New York: Wiley.

Hanson, J.B., 1984. The functions of calcium in plant nutrition. *Advances in Plant Nutrition.* 1: 149-208.

Hewitt, E.J. and Smith, T.A. 1974. *Plant Mineral Nutrition.* London: English Universities Press.

Mengel, K, and Kirkby, E.A. 1979. *Principles of Plant Nutrition.* Berne: International Potash Institute.

Wald, G. 1962. Life in the second or third periods; or why phosphorus and sulfur for high-energy bonds. In *Horizons in Biochemistry*, eds. M. Kasha and B. Pullman, pp. 127-42. New York: Academic Press.

9

Genotypic Differences in Plant Nutrition

Four critical resources limit global food production, namely, land, water, energy and fertilizers. In the past, up to the middle of the present century, most of the yearly increases in agricultural productivity came from expanding the land area under cultivation. In the developed countries it is unlikely that further gains can be obtained through this strategy. In fact there is increasing concern among soil scientists regarding the reverse trends. For example, in Canada approximately 43 million hectares of land are presently under cultivation. In the period from 1971 to 1976, (for which detailed figures are available), the conversion of high quality agricultural land to non-agricultural uses amounted to 38,211 hectares. Most of these losses were in the climatically favourable areas of Ontario and Quebec. In addition, a recent assessment of the state of agricultural land in Canada identifies extensive deterioration due to erosion (by wind and water), fertility reduction (resulting from nutrient removal by crops without adequate fertilizer application), acidification (due to nitrogenous fertilizers and acid rain), salinization (due to irrigation practices which often concentrate salt in the rooting zone) and other causes. Such gloomy pronouncements are not unique to Canada.

Similar problems, to varying degrees, beset most of the world's agricultural lands.

Yet, despite these problems, global food production has increased substantially. In the period from 1971 to 1983, for example, total production of the major food crops, wheat, rice, corn and potatoes rose 1.3 times from 1.19 to 1.58 x 10^{12} kg per annum. Such is the impact of increased productivity, that European farmers are currently being encouraged to reduce lands under cultivation by 15%, and costly subsidies to farmers (to compensate for low grain prices) have become a major political problem in developed countries. Such improved yields have almost certainly come about largely through an intensification of cultivation methods and through improved strains of the major crop plants. One indication of this intensification is to be found in the global pattern of fertilizer use which has climbed from 6.9 x 10^{10} kg per annum in 1970-1971 to 12.5 x 10^{10} kg in the period from 1983 to 1984. Not surprisingly, this has added considerably to the cost of producing our food. For example, if the dollar equivalence of revenues arising from crops in Canada in 1971 is set at 100 and the cost of fertilizer at this time is also indexed at 100 then in 1983 the corresponding figures have risen to 295 for revenues and 310 for fertilizer costs. One side-effect of the deterioration in land quality, referred to above, is that even greater quantities of fertilizers or soil amendments (e.g. lime) will be required even to sustain yield. In this regard a comparison of the annual fertilizer use for the twelve year period 1971-1983 shows that this increased by a factor of 1.8 while crop yields (see above) rose by a factor of only 1.3. Predictions of fertilizer requirements by the year 2000, from the United Nations Industrial Development Organization, suggest that demand for N will have risen to $\sim 16 \times 10^{10}$ kg, from a present 6×10^{10} kg. For P and K, requirements are anticipated to increase approximately two to threefold to about 7.0×10^{10} kg. While the best estimates appear to indicate that world reserves of P and K ores are quite adequate to supply demands for at least 2-300 years, one cannot doubt that the cost in energy terms of mining and refining these ores will be substantial. For the case of N fertilizers, greater than 95% are produced from nitrogen gas (via the Haber Process) by heating together N_2 and H_2 at high temperature and pressure to generate NH_3. The majority of world NH_3 production ($\sim 70\%$) is based upon natural gas (to generate the required H_2) with relatively little use of coal or oil. Although the world reserves of natural gas appear to be adequate to supply the increased demands for N fertilizer, concerns have been expressed for North America because of the high rate of fertilizer use and the relatively small ($\sim$ 10%) proportion of world natural gas reserves located in this continent. More importantly, however, it is the energy intensive nature of the synthetic process for NH_3 generation which is cause for concern. On average, production of a ton of N fertilizer requires six and sixteen times more energy, respectively, than is required to produce an equivalent of P and K fertilizer.

Perhaps, not surprisingly then, increasing numbers of scientists are placing greater emphasis upon the selection of improved plant genotypes to solve the long-term problems of food production, particularly for the developing nations. This does not simply refer to the generation of strains which can increase yield under the 'ideal' conditions of high fertilizer application. Some critics of the Green Revolution have noted that the fertilizer demands of the high yielding varieties represent too great a cost burden for small farmers in developing nations. Rather, it includes the generation of strains which may be capable of exploiting marginally arable lands; varieties which can tolerate acid soils, saline soils, drought stress and temperature extremes. In addition, it may involve selection for greater efficiency of utilization of the absorbed nutrients so that yields may be sustained at lower cost. In this chapter we consider genotypic differences in inorganic nutrition, emphasizing differential responses to nutrient availability and possible causes for these differences. In addition, we considered differential responses to environmental stress.

9.1 Genotypic Differences in Mineral Nutrition

Plant species occupy a diverse range of habitats. Those vary from the relatively moderate environments of temperate zones to the harsh extremes of arid deserts and arctic tundra. Even in temperate zones, localized variations in soil chemistry, as in the serpentine soils (Section 1.2), may impose such severe constraints that few species can survive. Where the environment is more hospitable, species may be so numerous that competition for essential resources (e.g. water and inorganic ions) may be quite intense, both between and within plant species.

Plants have evolved numerous adaptations to ensure survival under these conditions. In so far as plant nutrition is concerned, these involve both morphological and physiological characteristics. In the preceding chapters we have dealt with many of these adaptations, particular emphasizing features which plants possess in common. In this section, by contrast, we consider variations in these traits, which appear to enable plants to compete more effectively and survive the limitations of their habitats.

It has long been recognized that individual species may differ both qualitatively and quantatively with regard to their requirements for inorganic elements. While the need for the group of sixteen elements is universal, particular species require additional elements. In Section 8.1A, for example, it was noted that elements such as Si and Na are required by some plant groups.

In addition it is known that requirements for elements such as Co or Ni may vary according to the mode of nutrition of the plant. For instance, Co is required by N-fixing bacteria. This element is therefore a requirement for legume/Rhizobium associations and strong growth responses have been obtained by the addition of Co to nutrient media supplied to legumes (such as soya beans) depending upon nodule activity for their N supply. Nickel is considered to be a cofactor for the enzyme urease which breaks down urea to NH_3 and CO_2. Hence this element is required under conditions in which the bulk of a plant's N source comes from urea. Neither of the above elements can satisfy the criteria for essentiality described in Chapter 8, because their function can be replaced by means of an alternative N supply. Nevertheless they do illustrate the qualitative differences which may exist under special circumstances among species.

Of perhaps greater interest to physiologists and to ecologists are the quantitative differences in requirements for inorganic elements among, and even within, species. Some of the early fertilizer trials demonstrated clear-cut examples of differential responses to applications of N, P and K in field experiments. In Russell's text "Soil Conditions and Plant Growth", for example, data are collected demonstrating N responses of several field crops. Members of the cabbage family appear to give much stronger responses to applied N (at 31 kg ha^{-1}) than do potatoes or cereals. Here it is important to consider which portion (leaves, roots, tubers, seeds) will constitute the harvest. Clearly, where leafy material represents the harvestable portion of the plant it might be anticipated that a strong N response would be evident. One of the earliest controlled studies, using hydroponic facilities to investigate species differences in the accumulation of cations, was that of Collander (1941), referred to in Chapter 8. Using 21 different species, Collander established that the greatest differences among species existed for Na and Mn accumulation. Expressing the accumulation of each cation as a percent of total cation content it was observed that the value for Na ranged from 29 percent for *Plantago maritima* (a halophyte) to 0.9 percent for *Fagopyrum* (buckwheat). Differences for Mg and Ca were not as large but still substantial.

Calcium accumulation ranged from 33 percent of total cation in *Fagopyrum* to 8 percent in *Avena* (oat). For Mg, equivalent maximum and minimum values were 39 percent for *Atriplex* and 11 percent for *Plantago maritima*. The smallest variations were observed for K, with *Avena* showing the highest proportion (73 percent) and *Fagopyrum* the lowest (39 percent). Several generalizations were illustrated in this study. The legumes and herbaceous dicots tended to have higher Ca levels than the monocots, while monocots tended to have higher K levels than the dicots. It was also apparent that halophytes accumulated greater quantities of Na than other species. It should be stressed that all species were exposed to identical growth media, so that the differences in patterns of accumulation were due to inherent (genetically determined) differences.

At the Rothamsted Experimental Station in England, fertilizer effects on plant growth and diversity in permanent grasslands have been investigated for over a hundred years. These experiments have provided a valuable testimony to the influence of soil chemistry upon species composition of a community. Where experimental plots have received no fertilizer applications there is great species diversity ($\sim$ 60 higher plants), with no clearly dominant species and a consistently low yield of hay. Applications of various fertilizer combinations have brought about substantial changes in the species composition of the treated plots. Yields are generally increased but species diversity has often declined and particular species are seen to flourish. The conclusion is inescapable that species not only respond differentially to the availability of the nutrient elements but their responses may be sufficiently different that the community structure is dramatically changed. Unfortunately in such experiments the effects of fertilizer applications are sometimes complicated by concomitant changes of other factors such as soil pH. Nevertheless, more carefully controlled experiments of numerous ecophysiologists have confirmed the differential nature of the responses to nutrient availability in many wild and cultivated species.

For example, Bradshaw and his collaborators have investigated the responses of several grass species to the availability of the major nutrients. In response to NO_3^- availability, for example, the dry weights of certain species increased by factors of 10 to 20 fold in sand culture (Fig. 9.1). However, it is also evident that there was enormous variation in the response to applied N. Growth of *Nardus stricta* (mat-grass), for example, was strongly inhibited at the higher levels of NO_3^- availability. When experiments of this sort have been repeated using different plant species and varying the rates of application of K, P or Ca, essentially similar results were obtained.

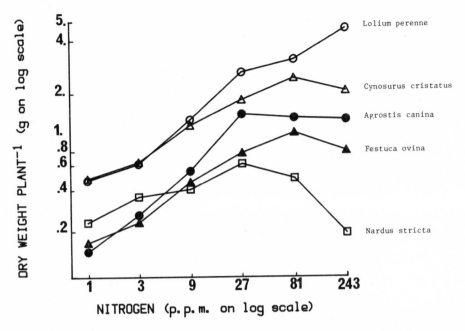

Figure 9.1. Variation in dry weights of 5 grass species in sand culture, in relation to variation in level of NO_3^-. (From Bradshaw *et al. Journal of Ecology*, 52: 665-676, 1964)

In many of these studies, authors have sought to interpret the characteristic associations between particular plants and the soils they normally occupy by reference to the results of experiments conducted under controlled conditions. Typically, species from nutrient-poor habitats grow satisfactorily (though slowly) at low levels of nutrient availability and fail to respond strongly to increased nutrient levels. As in the case of *Nardus* (Fig. 9.1), growth may actually be depressed at high N availability. The strategy of slow growth is now widely considered to represent an important mechanism for exploiting infertile habitats. By contrast, species from more fertile habitats demonstrate strong positive responses to increasing nutrient availability, and, particularly at higher nutrient concentrations, grow much more rapidly than species from the nutrient-poor habitats.

Among the agriculturally important crop species it has long been appreciated that differences occur, not only in various aspects of plant nutrition at the species level (see above), but also at the variety or cultivar level. An important review of this topic by Vose (1963), then at the Welsh Plant Breeding Station in Aberystwyth (North

Wales), records a voluminous literative dealing with varietal differences in yield responses to P, N and K. Much of the work deals with cereal crops, and some of these studies were undertaken as far back as the 1920's.

Clearly, differences in yield response may arise from several sources. These might include differences in rates of absorption of limiting nutrients, patterns of allocation of the absorbed nutrients and the differences in efficiency with which the absorbed nutrients are used in metabolic processes.

9.2 Yield Curves

When the yield of a particular crop species is plotted against the level of fertilizer application or nutrient availability, the resulting plot commonly resembles a rectangular hyperbola. For example, in Figure 9.2

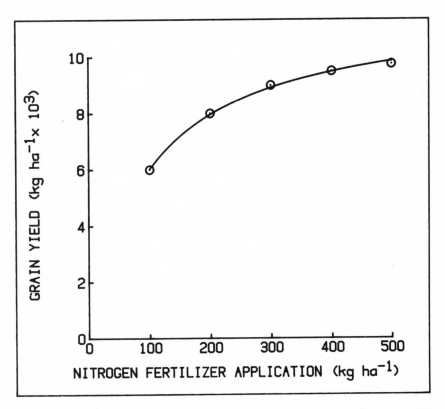

Figure 9.2.   Representative yield curve for a wheat crop as a function of N fertilizer application.

a plot of wheat yield (Y) is shown as a function of N fertilizer additions (S). As fertilizer additions are increased in the lower concentration range (0-100 kg ha^{-1}) there is an almost linear yield response and $\frac{dY}{dS}$, the increment of yield for an increment of fertilizer, is at its highest value (Table 9.1). Although absolute yield increases with further additions of fertilizer (> 100 kg ha^{-1}) it is clear that we are into a situation of diminishing returns, as shown by the declining value of $\frac{dY}{dS}$. Consider the situation where N fertilizer costs 0.73 cents kg^{-1} and the cash value of grain is 23.5 cents kg^{-1}. Without even considering all the other costs involved in producing the crop, Table 9.1 shows that the system is most profitable at a N application rate of 400 kg.ha^{-1}. Beyond this point the small increment of yield does not warrant the further expense of fertilizer, and profit ha^{-1} actually declines.

Yield curves for some plant species correspond reasonably well to rectangular hyperbolae (as in the above example) so that yield can be predicted by means of equations similar to the Michaelis Menten equation. For instance, we might define a maximum yield (Y_{max}: analagous to V_{max}) and a half-saturation concentration ($C_{0.5}$: equivalent to K_m). Yield at any fertilizer application rate would then be given by the following equation:

$$Y = \frac{Y_{max} \cdot S}{C_{0.5} + S} \tag{9.1}$$

Calculating Y_{max} and $C_{0.5}$ from the data of Table 9.1 gives values of 11.73×10^3 kg h^{-1} for maximum yield and 94.56 kg ha^{-1} for $C_{0.5}$

Table 9.1. Yield, increment of yield per increment of fertilizer ($\frac{dY}{dS}$), cash value of cereal crop, cost of fertilizer and profit as a function of the rate of fertilizer application, (see Figure 8.1 and text).

Nitrogen Fertilizer Application Rate (S) kg ha^{-1}	Yield (Y) kg ha^{-1}	$\frac{dY}{dS}$	Cash value of crop ha^{-1} at 23.5 cents kg^{-1}	Fertilizer cost ha^{-1} at 0.73 cents kg^{-1}	Profit ha^{-1}
			A	B	A-B
100	6000	—	$1410	$ 73	$1337
200	8000	20	$1880	$146	$1734
300	9000	10	$2115	$219	$1896
400	9500	5	$2232	$292	$1940
500	9750	2.5	$2291	$365	$1926

(the fertilizer application required to give 50 percent of maximum yield). Substituting these values into Equation 9.1 for a N application rate of 200 kg N fertilizer ha^{-1} predicts a yield of 7.96×10^3 kg ha^{-1}.

Another well known equation used for predicting yield is the Mitscherlich equation:

$$\frac{dY}{dS} = k\,(Y_{max} - Y) \qquad (9.2)$$

where k is a constant for the system under investigation. Clearly, this equation indicates that $\frac{dY}{dS}$ will be largest when yield Y is far from the maximum yield (Y_{max}) as was evident in Table 9.1. As Y approaches Y_{max}, ($Y_{max} - Y$) approaches zero; consequently $\frac{dY}{dS}$ is reduced toward zero. Although yield curves may demonstrate good approximations to the above equations, for many species there may be considerable deviation from these idealized representations. For example, growth may be inhibited at excessive levels of P or N and the growth of plants from nutrient-poor environments may show positive responses only over a limited concentration range. Figure 9.3 (Asher and Loneragan, 1967) shows examples of growth curves for several plant species in response to external Pi concentrations, maintained at constant levels by continuous-flow culture methods. It is immediately

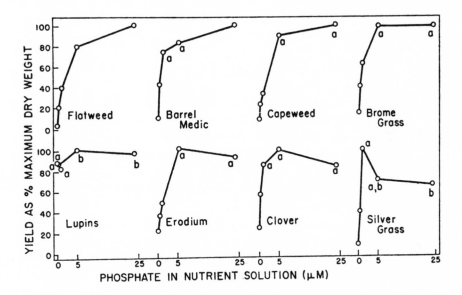

Figure 9.3. Yields of 8 plant species (as percent of maximum dry weight) grown in solution culture as a function of available Pi concentration. Maximum yield for lupin was 732.9 mg plant^{-1} and for silver grass was 66.5 mg plant^{-1}. (From Asher and Loneragan, *Soil Science*, 103: 225-233, 1967)

apparent that the different species have quite different Y_{max} and $C_{0.5}$ values. Furthermore, the ambient concentrations required to obtain optimum growth ranged from 1 μM for silver grass to 25 μM or greater for flatweed and barrel medic.

Experiments conducted at constant ambient concentrations, as in the above cases, provide plants at steady-state under controlled conditions. This is essential for comparisons of genotypic differences of the sort explored by Asher and Loneragan (Fig. 9.3). Under field conditions, however, the situation may be quite different. The continuous replenishment of nutrient in the laboratory system means that an infinite supply of the particular nutrient is available. If the plant can acclimate its transport system to obtain the requisite uptake rate at low external concentration then the Y_{max} can be achieved. Thus, growth is limited by the extent to which the plant can acclimate to ambient concentration. In the field, however, there is only a finite resource available whose delivery to the root may be limited by diffusion (Section 3.3b). There are also interspecific competitors for the resource (weeds) as well as intraspecific competition (due to members of the same species or crop). The intensity of this competition will be a function of factors such as planting density and the magnitude of the resource which is available. Nevertheless, in soils, as in flowing solution culture, the capacity for growth (and yield) has repeatedly been reported to vary among different species when grown under identical conditions.

9.3   Variations in Nutrient Uptake

In seeking to explain genotypic differences in response to nutrient applications crop scientists and plant breeders have undertaken studies of rates of nutrient accumulation by crop species since the 1920's. In many of these studies hybrids generated from specific crosses were included in the evaluation so that the inheritance of patterns of nutrient accumulation could be evaluated. Usually, however, rates of absorption of particular elements were not directly determined, but based upon an elemental analysis of the tissue at the end of growth periods which sometimes lasted for several weeks. Nevertheless, such analyses revealed considerable differences in tissue concentrations of N, S, P, K, Na, Mg, Ca, Mn, Zn, Fe, and Cu among varieties of crop plants. Clearly, in the absence of more detail it is difficult to evaluate whether such differences arose because of morphological differences among roots of the varieties or physiological factors such as differences of net uptake (per unit weight of root), partitioning between root and shoot, or even growth rates (since growth will tend to dilute absorbed nutrients).

In more recent years physiologists have undertaken more carefully controlled experiments, often measuring ion influx over relatively short periods by means of isotopically labeled solutions. Epstein and Jefferies (1964) reported that the affinities (K_m values) for K^+, Rb^+, and Na^+ uptake by roots of barley differed considerably among varieties. Similarly, Pettersson (1978) investigated eight barley varieties and also reported substantial differences in rates of uptake per unit root weight.

These interesting studies must be interpreted with caution. We have seen from Chapter 7 that growth rates can influence measured rates of uptake. When plant tissues approach their set point concentrations, uptake rates may be determined by growth rates. Thus it can be argued that varietal differences in uptake are not the causes, but rather the results, of differences in growth rates. Nevertheless, if we accept that these correlations exist, they may be usefully employed in breeding programmes to identify valuable genotypes at the seedling stage. Moreover, when care is taken to investigate rates of uptake under conditions where growth is nutrient-limited (e.g. in the fertilizer range below 100 kg N ha^{-1} in Figure 9.2) then rates of uptake should determine growth and differences in uptake rates among varieties may be quite important. For example, in my laboratory we have observed that some varieties which have high rates of K^+ absorption, even at low ambient concentrations, acquire greater proportions of available K^+ and reach their potential growth rates at much lower levels of available K^+ than varieties characterized by low rates of uptake. Table 9.2 shows K^+ influx for 13 varieties of barley grown at 60 μM K^+. Influx values were also measured at this concentration. These results demonstrate the wide range of influx values for plants grown under identical conditions. These varieties also demonstrated considerable variations in growth rates. Varieties with high fluxes of K^+ tended to have higher growth rates, but in some cases growth rates and K^+ influx values were poorly correlated for reasons to be discussed in Section 9.4.

Recently we have been investigating the growth of these barleys in competition with weed species such as wild oats (*Avena fatua*). Fergus, a barley variety which has high rates of K absorption (Table 9.2) was an extremely good competitor and reduced the weed growth to a greater extent than the weed reduced the growth of barley (Figure 9.4).

Table 9.2. K^+ influxes values for roots and total plant fresh weights of 13 barley varieties. Plants were grown at 60 μM K^+ plus other inorganic nutrients and fluxes were measured at the same concentration during a 10 min influx period. (From Siddiqi and Glass, *Can.J. Bot.* 61:1551-1558, 1983)

Variety	Influx ($\mu mol\ g^{-1}\ h^{-1}$)	Plant fresh weight ($g\ plant^{-1}$)
Steptoe	5.97	2.25
Fergus	5.22	3.26
Hector	5.08	1.74
Vantage	4.93	1.32
Betzes	4.69	2.36
Paragon	4.52	2.14
Fairfield	4.25	2.30
Bonanza	4.05	1.13
Keystone	4.04	1.58
Himalaya	3.97	1.64
Klondike	3.94	2.06
Olli	3.88	1.22
Conquest	3.43	0.85

Plants were grown in pots of sand irrigated with nutrient solution either alone (3 plants per pot) or with a competitor (3 + 3 plants per pot). The dry weight of Fergus (g plant^{-1}) was reduced by 30 and 32 percent, respectively at high and low K^+ levels whereas the growth of wild oats was reduced by 54 and 63%, respectively. It would seem that the high rates of K^+ absorption by Fergus, particularly at the lower level of K^+ supply is partly responsible for its competitive advantage. Growth of another barley, Bonanza (Table 9.2), which consistently demonstrates lower rates of K^+ absorption, was a poor competitor with the weed. Under the same experimental conditions, dry weights for this variety were reduced by 48 and 52 percent, respectively, in competition with the weed, whereas the weed's growth was reduced by only 39 and 20 percent, respectively.

Clearly the ability to obtain a limiting resource is only one component of the competitive character of a species. Another extremely important trait is the efficiency of utilization of the nutrient within the plant tissue.

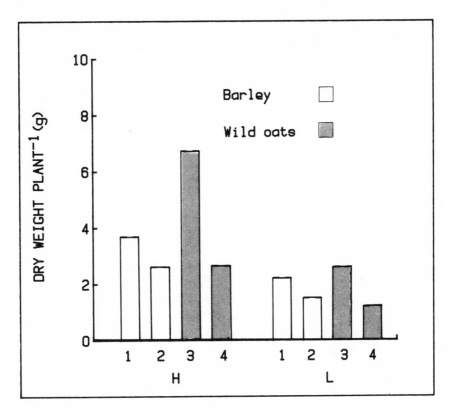

Figure 9.4. Dry weights of a barley variety (Fergus) and a wild oats strain (CS4O) under various growth conditions. Symbols: H — high K^+ supply; L — low K^+ supply; 1 — barley alone (3 plants pot^{-1}); 2 — barley + weed (3 + 3 plants pot^{-1}); 3 — weed alone (3 plants pot^{-1}); and 4 — weed + barley (3 + 3 plants pot^{-1}). (Data from Siddiqi *et al.*, *Annals of Botany*, 56: 1-7, 1985)

9.4 Utilization Efficiency

We commonly express the tissue concentration of a particular element, such as P or N in terms of μmol $g.f.w^{-1}$ or μmol $g.d.w^{-1}$. That is, the total element content of an organ or whole plant is divided by tissue weight to give a measure of concentration. If, instead, we divide tissue weight by the element content (which is equivalent to the reciprocal of concentration) we will obtain a measure of the plant mass per unit of absorbed nutrient. This quotient has been termed the efficiency ratio or utilization index and has been used to assess the efficiency with which a nutrient is used in the growth process. Thus, if several species or varieties are provided with an identical inorganic

resource in separate growth systems (such as sand culture or hydroponic facilities), then provided all (or most) of the resource is removed
by the plant, the growth response will provide a measure of the
efficiency with which the nutrient has been used for growth.

Numerous investigations have shown that varieties differ substantially in efficiency ratios. Table 9.3 shows data published by Gerloff
and his colleagues at the University of Wisconsin for varieties of
snapbean and tomato. Efficient strains gave efficiency ratios which
were substantially larger than those of inefficient strains. As the
availability of a growth-limiting nutrient increases it is commonly
observed that growth increases, but so also does tissue concentration
of the nutrient. Generally, plots of growth versus external ion concentration (where they demonstrate a reasonable approximation to rectangular hyperbolae) reveal lower half-saturation values ($C_{0.5}$) than
do the equivalent plots for tissue concentration of the nutrient. As a
consequence, at moderate nutrient levels growth is not increased to
the same extent as tissue concentration and thus the efficiency ratio
tends to decline. Figure 9.5 shows data for the barley variety Fergus in
which the ratio declined from close to 500 mg (dry matter) mmol^{-1} of
K^+ at 1 μM available K^+ to below 100 mg.mmol^{-1} at 100 μM K^+.

Table 9.3. Yield (g dry weight) and efficiency ratios (g dry wt. g of absorbed element^{-1})
for varieties of snapbeans, and tomato representing the extremes in efficiency ratios for
N, P and K. (From Gerloff, in *Plant Adaptation to Mineral Stress in Problem Soils*,
M.J. Wright, ed. 1976, Cornell University Press)

Plant	Element in Limiting Supply	Amount of Limiting Element Provided	Variety Number	Dry Weight	Efficiency Ratio
Snapbean	K	11.3 mg plant^{-1}	63	6.00	157
			58	8.83	294
Tomato	K	5 mg plant^{-1}	94	0.95	173
			98	1.97	358
Snapbean	P	2 mg plant^{-1}	1	0.87	562
			11	1.50	671
Tomato	N	2 mg plant^{-1}	51	2.51	83
			62	2.71	88
			34	3.51	110
			63	3.62	118

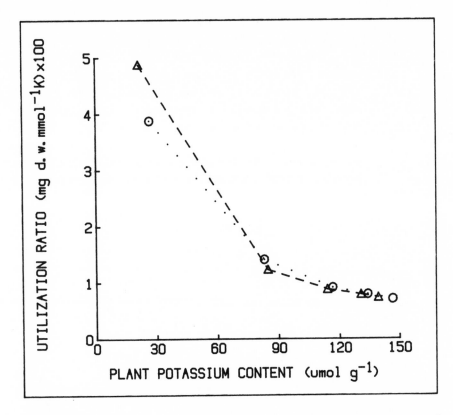

Figure 9.5.   Utilization ratios for two barley varieties Fergus (circles) and BT 334 (triangles) as a function of internal potassium concentration. (From Siddiqi and Glass, *Journal of Plant Nutrition*, 4: 289-302, 1981)

The factors responsible for the differences in utilization among different varieties have not yet been identified. Clearly, if experimenters express utilization in terms of total plant mass and total element content it may be that differential patterns of nutrient allocation between root and shoot may be responsible. Dunlop (1975) in New Zealand, reported that varieties of ryegrass differed in rates of K^+ translocation by their excised seminal roots. Subsequent experiments from the same laboratory have confirmed that there are also significant differences among different varieties in rates of translocation in nodal roots. Not only were differences identified among varieties of this species but, also, individual genotypes of a single variety ('Grasslands Paroa') were shown to differ substantially with regard to translocation rates. Both inter-varietal and intra-varietal differences in rates of K^+ uptake were also noted and translocation rates were found to be strongly correlated (r^2 values = 0.95 and 0.98, respectively) with rates of uptake (Fig. 9.6).

Other explanations for differential utilization rates may depend upon differences in the partitioning of nutrients at the sub-cellular level, e.g. between cytoplasm and vacuole. Generally, when nutrient supplies are adequate to sustain maximum growth rates, elements such as N, P, and K are stored in the vacuole, providing solute for the generation of turgor as well as a reservoir of essential elements (Chapter 6). When nutrients are in limited supply, cytoplasmic levels are held constant by removal of these elements from the vacuole.

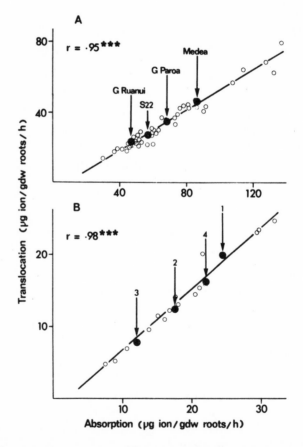

Figure 9.6. Translocation and absorption rates for K^+ by 4 cultivars of ryegrass (A) and four individuals of the variety Grasslands Paroa (B). Open circles represent individual values of measured fluxes, closed circles represent mean values. (From Dunlop and Tompkins, in *Transport and Transfer Processes in Plants*, I.F. Wardlaw and J.B. Passiovra, eds. 1976, Academic Press)

Turgor may now be provided by organic solutes such as sugars. Although the turgor function can be satisfied relatively non-specifically, cytoplasmic (metabolic) functions are much more specific. Differences in utilization may therefore depend on the extent to which the vacuole's reserve of inorganic nutrients can be removed to the cytoplasm and replaced by organic solutes. Techniques such as ^{31}P N.M.R. and X-ray microprobe analysis have confirmed the essential constancy of cytoplasmic ion concentrations in a limited number of higher plant species and could, theoretically, be applied to this particular problem.

Finally, there may be inherent biochemical differences among varieties in their responses to available nutrients. Recently Chapin and Bieleski (1982) have compared the responses to P availability of a wild barley (*Hordeum leporinum*) from low-P soil, with those of a cultivated variety (*H. vulgare*). Absorption rates of the two species were comparable, but the cultivated barley produced greater biomass (shoot weight, leaf size, shoot number etc.) at all levels of P supply (Fig. 9.7). The wild variety was not responsive to increased P supply,

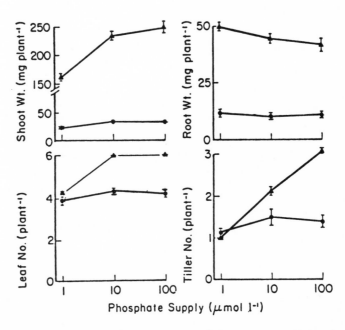

Figure 9.7. Shoot weights, root weights, leaf numbers of the the largest tillers, and tiller numbers of wild barley (•) and cultivated (▲) barley grown at 1, 10, and 100 μM Pi. (From Chapin and Bieleski, *Physiologia Plantarum*, 54: 309-317, 1982)

whereas the cultivated variety demonstrated a strong positive response. This lack of response to nutrient availability of wild species from infertile habitats was noted in the work of Bradshaw (Fig. 9.1). The widely held belief in the "slow-growth strategy" as the means of surviving in nutrient-poor habitats clearly requires explanation at the mechanistic level. What is the mechanistic explanation for the fact that when rates of absorption are so similar the wild species store absorbed nutrient whereas the cultivated species show accelerated growth? The answers to such questions await further study.

In my laboratory we have explored differences in K^+ utilization in barley varieties. Using the enzyme pyruvate kinase (an enzyme in the glycolysis pathway which is activated by K^+) we evaluated the concentration-dependence of activation. Could it be that K^+-efficient varieties required a lower $[K^+]$ for activation? The answer to this question reminded us of Huxley's famous remark, "The great tragedy of Science - the slaying of a beautiful hypothesis by an ugly fact". We discovered that there were only very small differences in the concentration of $[K^+]$ required for activation of pyruvate kinase among our varieties. Moreover, the concentrations required for maximum activation of the enzyme (*in vitro*) were about 4 mM, very much lower than current estimates ($\sim 100 - 150$ mM) for cytoplasmic $[K^+]$ (Fig. 9.8).

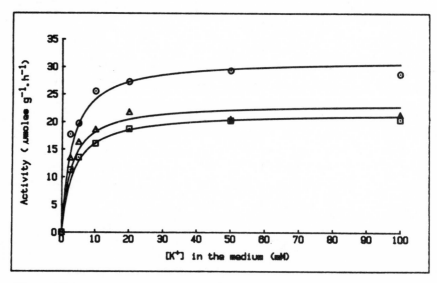

Figure 9.8. Pyruvate kinase activity of leaves of 3 barley varieties, Compana (o); BT 334 ($\triangle$) and Fergus ($\square$) as a function of K^+ concentration of the incubation medium. (From Memon, Siddiqi and Glass, *Journal of Experimental Botany*, 36: 79-90, 1985)

We observed that an inefficient variety (Betzes) was unable to transfer vacuolar K^+ to the cytoplasm when its vacuolar $[K^+]$ fell below 33mM. Even though average tissue $[K^+]$ was as high, or even higher than in efficient varieties, Betzes failed to maintain cytoplasmic $[K^+]$ and displayed symptoms which were typical of K^+ deficiency. By contrast an efficient variety (Compana) was able to reduce vacuolar $[K^+]$ to 20 mM and maintain its cytoplasmic K^+ concentration by transferring K^+ from the vacuole. In addition, Compana appeared to have a lower cytoplasmic requirement for K^+ (see Table 7.5, page 139) and maintained higher rates of protein synthesis than Betzes under these conditions.

There are now a large number of reports of differences in the uptake and utilization of inorganic nutrients among crop varieties. It is important that these differences be accounted for at the mechanistic level if systematic selections for such desirable physiological traits are ever to be incorporated into plant breeding programmes.

9.5 Genotypic Variation with Respect to Soil Toxicity Problems

It has been stressed that concentrations of the essential elements may exhibit seasonal and distributional variations in most soils (Section 3.1b). Up to this point we have emphasized the physiological and developmental adaptations which enable plants to procure sufficient quantities of these nutrients when they are in limited supply. However concentrations of these elements may also be sufficiently high in certain soils that toxicity problems arise for species not normally grown in these areas, or varieties which lack tolerance mechanisms. Examples include toxic effects of Mn, Fe, Cu, and even N (in the form of NH_4^+). Similarly, concentrations of non-essential elements such as Al, Na and, more rarely, Se present problems to particular genotypes. On a global basis, soils containing harmful levels of minerals (excluding saline and alkaline soils) amount to about 3×10^9 ha (~ 23 percent of the total land area of the world).

Another source of toxicity is provided via industrial pollution, involving both essential and non-essential elements. For example mine spoils containing excessive levels of Zn, Cu, Ni and other elements may present severe problems. Particularly where plants form part of a human food chain there is considerable interest attached to the absorption (or exclusion) of these elements and their detoxication. Only in recent times have we become aware of the existence of mechanisms conferring tolerance to toxic elements possessed by certain strains of plant species. The colonization of toxic mine spoils by such plants represents a fascinating example of natural selection at work; an example which is analogous to the often quoted industrial melanism in the peppered moth (*Biston betularia*). The problems of toxic soils, the underlying causes of toxicity and mechanisms of tolerance represent extremely large and complicated subject areas. In this section we will limit consideration of toxicity problems to situations associated with excessive levels of Al in acidic soils and to salt accumulation in saline and alkaline soils.

9.5a ACID TOLERANCE

Soils which are strongly acidic or strongly alkaline present considerable difficulties for the growth of most plant species. At extremes of pH ($>$ pH 9 and $<$ pH 3) these effects may be a direct consequence of pH *per se*. However, even at moderately acidic pH (6-4) there may be considerable reductions of yield which are due to indirect effects of pH. Soil pH can exert a potent influence upon the solubilities of both essential (e.g. Mn, and Fe) and non-essential (e.g. Al) elements. Al and Mn solubilities increase markedly below pH 5.0 bringing greater concentrations of these ions into soil solution. Under hydroponic conditions as little as 10 p.p.m. Al^{3+} ($\sim$ 0.37 mM) causes irreversible damage to apical meristems of susceptible roots. Plant breeders have generated many strains of cereals and other crop species which are capable of good growth in acid soils. These are referred to as "acid tolerant". However, as described above, the harmful pH effects are indirect; a feature of the presence of toxic elements in soil. In hydroponic facilities most species can tolerate low pH so long as these elements are absent. Thus, rather than refer to strains as acid tolerant they should perhaps be called Al or Mn tolerant, as the case may be. Among crop plants, legumes such as alfalfa and clover are particularly sensitive to acidic soils and growth reductions are evident below pH 6. One solution to this problem is liming (with CaO) which can dramatically increase soil pH. Yields of alfalfa, with and without lime additions, at different initial soil pH values are shown in Figure 9.9. In the Peace River region of British Columbia and Alberta, Agriculture Canada scientists recently investigated the yields of 18 barley varieties, with and without lime. In the Silver Valley, where soil pH was 4.5 (increased to 6.2 by liming) the average yield was 55 kg ha^{-1} without lime, increasing to 2411 kg ha^{-1} with lime. At the other extreme, species such as the tea plant (*Thea sinensis*) accumulate large quantities of Al throughout their lives, storing up to 4500 p.p.m. (4.5 mg g

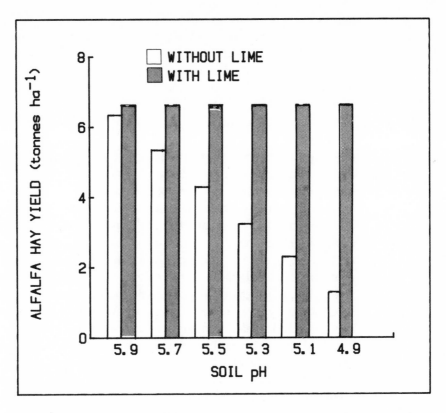

Figure 9.9. Yields of alfalfa as a function of original soil pH, with or without lime. (From Agriculture Canada Publication 1521, 1974)

d.w.$^{-1}$) within older leaves. X-ray microprobe analysis has indicated that much of this element is concentrated within the walls of epidermal and mesophyll cells. Why such large quantities are accumulated (even when exchangeable Al levels in soils are low), and why a toxicity problem is not encountered by these plants, is a mystery.

The cause(s) of the toxicity symptoms resulting from excessive levels of Al are not entirely clear. In fact there may be several deleterious effects of this element, including the inhibition of DNA synthesis at root apical meristems, the "fixing" of Pi in soil and root by conversion to unavailable form, as well as the inhibition of various enzyme activities including those involved in respiration and nutrient absorption.

The mechanisms responsible for the observed "acid tolerance" of species and varieties is also variable. In some cases it appears that Al is excluded from the tissues so that tolerant genotypes contain smaller quantities of this element than susceptible varieties. Foy at the Plant Stress Laboratory (U.S.D.A. at Beltsville, Maryland) has shown that Al-tolerant varieties of several crop species are able to increase the pH of their external media, hence reducing the concentration of Al in solution. This phenomenon has been confirmed in soil samples adjacent to roots of tolerant strains. Changes of pH in the external media arising from differential ion absorption (anion in excess of cation absorption) were considered in Sections 5.3 and 7.3. In other cases tolerant varieties may accumulate greater quantities of Al than susceptible varieties so that tolerance appears to be a function of the internal detoxication in these genotypes. For example, in azaleas, cranberry and rice, roots serve to sequester Al, and shoot Al levels are kept low. In this regard the chelating properties of organic acids (oxalate and citrate), or even phenolic compounds, may be important. A likely site for this localization of Al is within the vacuoles of root cells. A third, rather large category of plants, accumulates considerable quantities of Al within older leaves but maintains quite low levels within young leaves. This group, referred to as Al accumulator species includes tea and many tree species (e.g. pines).

Acid soils frequently contain high levels of Mn in addition to Al and, as in the case of Al, there is considerable variation among species and varieties in the capacity to tolerate excessive levels of this element. Tea, Japanese holly (*Ilex crenata*) and a Japanese tree species *Acanthopanax sciadophylloides* have been shown to concentrate Mn within leaf tissue. Recent evidence suggests that Mn in such plants may be chelated with oxalic acid within vacuoles of older leaves.

9.5b SALT TOLERANCE

Soils of many regions of the world contain excessively high concentrations of inorganic salts which cause severe growth reductions in numerous crop species. Current estimates suggest that a total of about 1×10^9 ha may be involved, accounting for almost 8 per cent of the world's total land area. Regions where such soils exist include U.S.A. (Californian serpentine soils), India, Pakistan, Egypt, U.S.S.R. and Holland. Soil scientists usually recognize two categories of salt affected soils. Saline soils contain high concentrations of the chlorides, sulfates and carbonates of Na, Ca and Mg and generally have pH values below 8.5. Alkaline soils contain high concentrations of sodium carbonates which generate a strongly alkaline ($>$ pH 8.5) reaction. Such soils are commonly found in arid or semi-arid regions where high rates of evaporation of unleached soils concentrate salts in the upper soil layers. Salinization can also result from irrigation practices where high rates of evaporation serve to concentrate salt in the rooting zone.

In California, salinization has been a problem ever since irrigation became a practise in the second half of the nineteenth century. Figure 9.10 provides dramatic illustration of the impact of salt build-up in cotton fields in the San Joaquim Valley, California.

The severe stunting of plant growth among sensitive species may arise as a direct consequence of salt and (particularly in alkaline soils) from indirect effects on soil texture. Alkaline soils commonly become pasty when wet, and heavily compacted when dry. As a consequence water and air movement is limited in such soils. Apart from these indirect effects on plant growth there are two more direct hazards to plants associated with high concentrations of salt. Firstly there is the osmotic stress. In order to maintain turgor under such conditions solute concentrations within plant tissues must be even more concentrated than in the external environment, to avoid desiccation. Halophytes, plants which normally occupy saline environments, usually maintain turgor by the accumulation of high concentrations of salts such as NaCl without deleterious effects. Several authors have proposed that high internal salt concentrations in such plants are tolerated by sequestering NaCl within the vacuoles. Under such conditions

Figure 9.10. Infrared photograph of a "salted-out" cotton field in San Joaquim Valley, California. The region to the upper right and lower portions of the photograph are almost bare of plants due to salt accumulation. (From Beck, *California Agriculture*, 36: 16-17, 1984)

the cytoplasmic phase is maintained isoosmotic with the vacuole by means of non-injurious "compatible" organic solutes. Examples of these solutes include glycerol, mannitol, sucrose, amino acids (particularly proline) and various other N-containing derivatives such as glycine-betaine (Fig. 9.11).

In addition to osmotic stress, plants in high salt environments suffer the potential hazards of specific ion toxicities due to excessive accumulations of ions such as Cl^-, SO_4^{2-}, Na^+ and Mg^{2+}. Clearly the halophytes have resolved this problem but in sensitive non-halophytes subjected to high levels of salt there appears to be an inability to sequester these salts within vacuoles, with disastrous consequences.

Genotypic differences among species in relation to salinity are pronounced. The diversity of marine organisms and the substantial

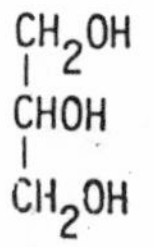

a. glycerol

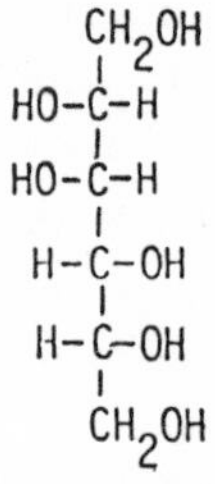

b. mannitol

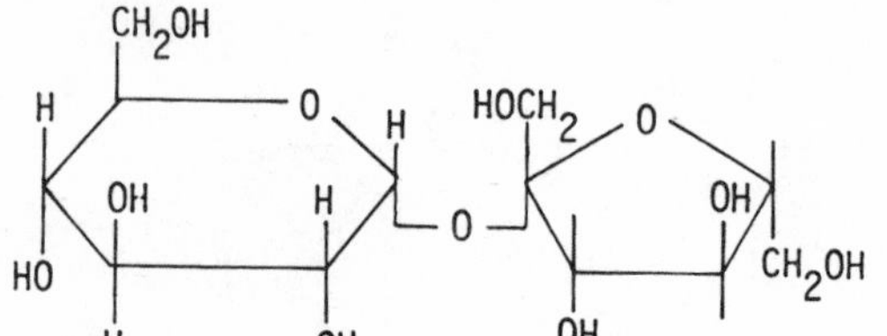

c. sucrose

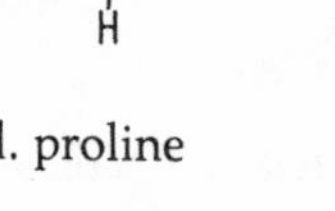

d. proline

e. glycine-betaine $(CH_3)_3 \overset{+}{N} CH_2 COO^-$

Figure 9.11. Structural formulae of various "compatible" organic solutes which may serve to reduce the water potential of the cytoplasmic compartment.

figures for global primary production in this environment (approximately one third of all photosynthetic carbon fixation) attests to the efficiency with which many plant species are adapted to saline habitats. By contrast, many species such as beans demonstrate as much as 50% yield reduction at 50 mM external NaCl concentration (sea water contains approximately 500 mM NaCl). Figure 9.12 shows the growth responses for several species ranging from a representative halophyte (*Suaeda maritima* - curve 1) which grows well at up to 500 mM external NaCl, to sensitive species such as beans (curve 4) which do not survive even to 100 mM NaCl.

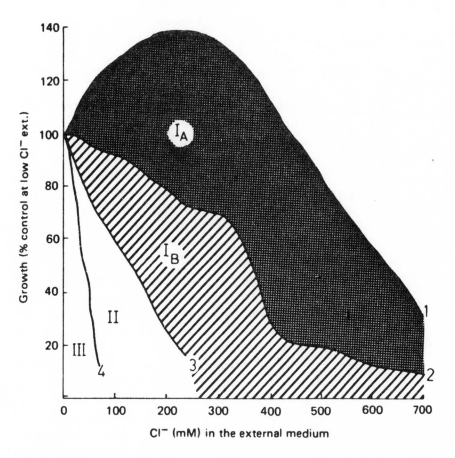

Figure 9.12.   Generalized groups of plants, ranging from halophytes (IA), which grow well at 200-500 mM NaCl, to very salt-sensitive nonhalophytes (III). The four lines are actual growth responses of *Suaeda maritima* (1); sugar beet (2); cotton (3); and beans (4). (From Greenway and Munns, *Annual Review of Plant Physiology*, 31: 149-190, 1980)

On a global basis the salinity problem for agriculture is immense and salinization of irrigated lands adds to the problem. Paradoxically, apart from the salinity problem, the affected lands are often particularly fertile in so far as their soils are unleached (because of the low rainfall) and their salinity has effectively discouraged colonization or agricultural use. The potential for reclamation of this land certainly exists, but it represents an extremely costly venture. Hence, there is considerable world wide interest in increasing salt tolerance among agricultural species. In any screening exercise where large numbers of individual plants are to be examined a convenient isolation technique is required. Fortunately, the deleterious effect of salt on survival of non-tolerant genotypes is a ready made selection criterion which has been used to great advantage by Epstein and his collaborators at the University of Davis, California. Using Composite Cross XXI, a genetically heterogenous line of barley (synthesized from 6200 lines), Epstein made use of NaCl stress to select a small group of survivors (99.7% failed to survive the treatment). These and other barley lines were tested at Bodega Marine Laboratory in California. Plants were grown in plots among the dunes and irrigated with sea water supplemented by N and P (Fig. 9.13). The outcome of this remarkable experiment was a yield of 1082 kg ha^{-1}, while subsequent trials have yielded closer to 1500 kg ha^{-1}. As a point of comparison, best yields of barley at Swift Current, Saskatchewan were reported for the variety Fairfield at 6,651 kg h1^{-1} in 1977. The three closest ranked barleys after Fairfield gave yields of 4650, 4045 and 3721 kg ha^{-1}, respectively. Compared to these values the Bodega yields are small. Nevertheless, considering the limited time during which these selections have been made, the potential for improvement is immense.

Another approach adopted by Epstein's group has involved incorporating salt tolerance from wild species of tomato (*Lycopersicon cheesemanii*) into varieties of the commercial species (*L. esculentum*). Seeds of the salt-tolerant *L. cheesemanii* were collected from the Galapagos Islands where plants grow close to the seashore. Unfortunately, the fruit of this species are little more than the size of a marble. Crosses between the two species have yielded some progeny which are moderately salt tolerant, with fruits the size of cherry tomatoes.

The two examples above provide dramatic evidence of the potential to improve salt tolerance by judicious use of existing germplasm and standard breeding techniques. There exists an enormous technical potential for improving strains by means of tissue culture and the emerging discipline of genetic engineering. It is to be hoped that concepts developed in these areas will bear fruit in the more practical areas of crop productivity.

Figure 9.13.   Salt tolerant strains of barley growing in the sand dunes at Bodega Marine Laboratory Station in California. Plants were irrigated with sea water supplemented with nitrogen and phosphorus. (From Epstein and Norlyn, *Science*, 197: 249-251, 1977)

Summary

Genotypes of wild species and varieties of cultivated plants demonstrate both qualitative and quantitative differences with respect to their elemental requirements. These genotypic variations may be the results of differences in rates of ion absorption, the allocation of absorbed nutrients between plant parts and/or in the efficiency with which absorbed nutrients are utilized.

Species and crop varieties also exhibit substantial differences in response to potentially toxic excesses of both essential and non-essential elements in the soil environment. The isolation of strains of cultivated crops which can tolerate the stresses imposed by these problem soils holds considerable promise for the exploitation of presently nonproductive lands.

Further Reading

Bradshaw, A.D., Chadwick, M.J., Jowett, D., and Snaydon, R.W. 1964. Experimental investigations into the mineral nutrition of several grass species. IV. Nitrogen level. *Journal of Ecology*, 52:665-76.

Chapin. F.S., III. 1980. The mineral nutrition of wild plants. *Annual Review of Ecology and Systematics*, 11:233-60.

Epstein, E., and R.L. Jeffries. 1964. The genetic basis of selective ion transport in plants. *Annual Review of Plant Physiology*, 15:169-84.

Epstein, E. 1976. Adaptations of crops to salinity. In *Plant Adaptations to Mineral Stress in Problem Soils*. New York: Cornell University Press.

Epstein, E. 1983. Tolerance of salinity and other mineral stresses. In *Better Food For Crops*. CIBA Foundation Symposium 97, London. Pitman.

Foy, C.D., Chaney, R.L., and White, M.C. 1978. The physiology of metal toxicity in plants. *Annual Review of Plant Physiology*, 29:511-66.

Gerloff, G.C. 1976. Plant efficiencies in the use of nitrogen, phosphorus and potassium. In *Plant Adaptation to Mineral Stress in Problem Soils*. New York: Cornell University Press.

Greenway, H., and Munns, R. 1980. Mechanisms of salt tolerance in nonhalophytes. *Annual Review of Plant Physiology*, 31:149-90.

Laüchli, A. 1976. Genotypic variation in transport. In *Encyclopedia of Plant Physiology*, New Series, Vol II.B. eds. U. Lüttge and M.G. Pitman, pp. 372-93. Berlin: Springer-Verlag.

Vose, P.B. 1963. Varietal differences in plant nutrition. *Herbage Abstracts*, 33:1-13.

Index

231